W9-AXG-102

SELLOUT

BY THE SAME AUTHOR

No Apparent Danger: The True Story of Volcanic Disaster at Galeras and Nevado del Ruiz

Hostage Nation: Colombia's Guerrilla Army and the Failed War on Drugs

SELLOUT

HOW WASHINGTON GAVE AWAY AMERICA'S TECHNOLOGICAL SOUL, AND ONE MAN'S FIGHT TO BRING IT HOME

VICTORIA BRUCE

B L O O M S B U R Y
NEW YORK • LONDON • OXFORD • NEW DELHI • SYDNEY

Bloomsbury USA
An imprint of Bloomsbury Publishing Plc

1385 Broadway	50 Bedford Square
New York	London
NY 10018	WC1B 3DP
USA	UK

www.bloomsbury.com

First published 2017

ISBN: HB: 978-1-63286-258-7
ePub: 978-1-63286-259-4

Library of Congress Cataloging-in-Publication Data is available.

2 4 6 8 10 9 7 5 3 1

Typeset by RefineCatch Limited, Bungay, Suffolk
Printed and bound in the U.S.A. by Berryville Graphics Inc., Berryville, Virginia

For
Dravin, Shale & Evelyn

CONTENTS

Prologue

In 1994, as an undergraduate in geology at the University of California, Riverside, I stood atop a rocky outcrop in the Mojave Desert. Our field trip guide—a geologist from the Molycorp mining company—handed me a chunk of light pink rock. "We were the world's first and only mine for rare earth elements," he explained. I had no idea what rare earths were, but apparently the metals in these rocks were of economic value and used in the latest tech products. The year before we were there, he told us, a group of Chinese geologists came. "We gave them a tour. We showed them everything about our operation—which type of rocks have rare earths, how we find them. Then the Chinese went home and found their own deposit a thousand times bigger." Now China was selling rare earths for next to nothing, and Mountain Pass was barely treading water—thanks to their own hospitality.

Sixteen years later, in 2010, the subject of rare earths exploded in the headlines. Apparently, China was now the *only* source in the world for the critical metals. What had become of Mountain Pass? China controlled not only all of the world's rare earth metals but the entire manufacturing chain for every goddamn thing made with them. How could America be so stupid?

I discovered the answer to that question over the course of hundreds of hours of intense investigation: folly, greed, hubris, and shortsightedness. I felt sick. I felt helpless. I felt like my daughter had no chance of growing up in the land of opportunity as I had. The most innovative country on earth, the country that put a man on the moon and split an atom, just sold everything off in return for cheap, disposable junk. I didn't want to write this book after all. It was just too dismal.

Then I met Jim Kennedy—a man with a burning mission to bring back manufacturing and innovation to America; a self-described stupid Irishman who won't take no for an answer, even when he's knocking on the Pentagon's door. The road he's traveled has cost him dearly, but I truly believe that no matter how many battles he loses, Jim will never give up this fight. And knowing that there are people like him in the world is enough impetus for me to tell a brutally hard story in the hope that one day, it may have a happy ending.

Victoria Bruce, June 2016
Riva, Maryland

1

Oak Ridge, Tennessee

On a January evening in 2011, in the lobby of a nondescript hotel off Route 62 in eastern Tennessee, Jim Kennedy opened his laptop and tabbed through a slide presentation. Hyper by nature, the native Missourian was coming undone. The next day was a big one. In the afternoon, he would give a talk that he knew would change the course of his life, and at the same time rescue the United States of America from the technological abyss that he believed it was hurtling toward. Kennedy had been to dozens of meetings with scientists over the last few years, but this was different. Oak Ridge National Laboratory had emerged from the hills as the largest single site of the Manhattan Project in 1943, becoming one of the greatest research institutions America had ever known. Some of the biggest brains on the planet appeared in its secret laboratories to fashion the most powerful energy release the world had ever seen. To Kennedy, the thought was mind-blowing. A self-described punk teenager who slept through five years of high school and who knew nothing about chemistry or nuclear physics just a few years back—was now invited to give a science and policy presentation to the current crop of Oak Ridge masterminds.

With Kennedy was John Kutsch, an engineer from Chicago with a built-in megaphone for a voice box who recently brought Kennedy into his inner circle of proponents for new, safer nuclear technology. Together, their mission was to convince Oak Ridge to restart its nuclear program using thorium instead of uranium for fuel, in a molten salt reactor rather than a solid fuel reactor—a design that successfully went critical in the Oak Ridge hills in 1965. The invention promised safe nuclear energy that couldn't explode because it wasn't cooled by pressurized steam. It also produced very little plutonium, mitigating the risk of nuclear weapons material making its way to countries that hadn't signed onto the 1970 Nuclear Nonproliferation Treaty—countries that included Pakistan, North Korea, Israel, South Sudan, and India. But after four years of continuous operation, the Molten Salt Reactor Experiment was unceremoniously abandoned by the Department of Energy, without a word of explanation to the Oak Ridge engineers.

In a matter of minutes, Kutsch went from confident about the following day's meeting to completely unnerved. "I started getting all these texts and e-mails about China, and I'm wondering what the hell is going on," he remembers. A barrage of expletives erupted from his giant mouth just as the head of the Oak Ridge nuclear energy department arrived in the hotel bar. Jess Gehin planned to meet Kennedy and Kutsch the next day, but he'd come by the hotel to greet the two, over beers, beforehand. Gehin, so mild-mannered and soft-spoken that he's barely audible at times, had something that he wanted them to see. Without the slightest dramatic lead-in, Gehin opened his laptop to an article published in the *South China Post* out of Hong Kong, along with a rough Google translation. The photo at the top showed a smooth-faced Chinese man with frameless glasses and graying temples: Dr. Jiang Mianheng, vice president of the Chinese National Academy of Science. "Yesterday, as the Chinese Academy of Sciences started the first of the Strategic Leader in Science and Technology Projects, '*The Future of Advanced Nuclear Fission Energy—Thorium Based Molten Salt Reactor System*,' was officially launched. The scientific goal is to use 20 years or so to develop a new generation of nuclear energy systems, all the technical level reached in the trial, and have all intellectual property rights."[1]

As bad as the translation was, the message was clear. "Our first reaction is like, 'Holy shit. We're fucked,'" says Kutsch.

The Chinese were going to develop the exact technology that he and Kennedy and were trying to get the United States to revive—technology invented right here in Oak Ridge and paid for by the U.S. government. Then China planned to patent the technology and sell the final product to the rest of the world.

The United States would be buying its own nuclear energy technology from China.

When Kennedy and Kutsch wondered aloud how China could be pursuing the same thing that Oak Ridge invented in the 1960s, Gehin told them something that rendered even Kutsch speechless. "He said there had recently been a bunch of high-level visits by Chinese scientists to Oak Ridge," Kutsch says. The head of the delegation was Jiang, the man in the article. Not only was Jiang one of the most prominent scientists in China but his ties to the Chinese government couldn't be any tighter; his father, Jiang Zemin (in China, last names come first), was the former president of the People's Republic of China. To Kennedy, the visit to Oak Ridge seemed more like an act of espionage than a scientific exchange of ideas, but Gehin seemed unconcerned and unforthcoming. "It was like an inside secret," said Kennedy. When pressed for details, Gehin downplayed the visit. "'It's no big deal. Don't panic. It's fine,' he told us."

Having grown up in a violent home, as the third of eight punching bags for a volatile Irish American father, Kennedy was quick to assess a threat and execute a survival maneuver. As with his early home life, there was really no bright side to look on. Still, he gave it his best shot. Having China in this race could be bad news, yes. But if Oak Ridge would get moving and be competitive, like in the old Cold War days, the United States could restart its own molten salt nuclear program and revolutionize safe nuclear energy. The U.S. would own the intellectual property and patents, and perhaps not spend the coming century dependent on buying Chinese nuclear reactors. Regardless of how Kennedy spun it in his head, knowing China had been right here in this national lab, ground zero of thorium molten salt technology, was a hard pill to swallow.

The next morning, the Chinese elephant that Jess Gehin introduced into the room followed Kennedy and Kutsch into the car and onto Route 62 toward the national laboratory. Kennedy drove the rental car past half-empty strip malls and abandoned buildings. Back in the mid-1940s, the town of Oak Ridge had a population of 75,000 and the feel of a bustling city.[2] But as the need for massive human resources to work in the lab plummeted in the years after World War II, the town began a unidirectional spiral downward. By the turn of the century, Oak Ridge was a place of abandoned malls and few gastronomic resources beyond fast food chains. Soon after arriving, spouses and partners of young researchers landing gigs at the laboratory began to count the days before they could leave. Grown kids of emeritus scientists rarely returned for a visit. Grandchildren living elsewhere in the country were seldom seen on the local playgrounds.

Part of the reason for the decay of modern Oak Ridge was exactly what brought the Army Corp of Engineers here in 1942. Just after being named Manhattan Project commander, Lieutenant General Leslie Groves chose the remote hills of this unpopulated region for a secret lab that would turn uranium from its natural state to fuel-grade, which could then be converted to plutonium for nuclear weapons. The scientists were cracking the most basic form of matter for first time in human history, and they knew they could be in for some colossal accidents. For Groves, the four high mountain ridges in the seventeen-mile-long valley offered a natural barrier to explosions. Another bonus was that the Tennessee Valley Authority just finished the nearby Norris dam project, and electricity came cheap.

At the lab's security entrance, an armed guard wearing a bulletproof vest checked Kennedy and Kutsch off a list of prescreened visitors. When they were let through the gate, part of Kennedy was still wondering why they would even let someone with his complete lack of scientific credentials into the national laboratory. As they passed the guard booth for the two-mile ride to the main campus, "All I'm thinking is, 'Some idiot just let me into Oak Ridge,'" he says.

Jess Gehin met them at the visitors' center with several other researchers. Their first stop was to tour the lab and the original Oak Ridge nuclear reactor, or "pile," so named because of the way the graphite blocks were piled on top of each other. The X-10 Graphite Reactor was the world's second man-made nuclear reactor, after Enrico Fermi's Chicago Pile 1. In only ten months, the reactor was designed and built and converting uranium-238 into intensely radioactive plutonium-239 for the first nuclear bombs. Kennedy and Kutsch were now able to get up close and personal with the X-10, which was now a museum display, with dusty mannequins inserting faux uranium fuel rods into a wall of graphite blocks.

The next stop on their tour was what they had really come to see: the Molten Salt Reactor Experiment. The underground reactor had been mothballed since 1970, but it was still a thing of beauty to Kutsch and Kennedy, who were just toddlers when it had "gone critical"—that is, begun its self-sustaining generation of power.[3] They were positive they could convince Oak Ridge management to bring this very reactor concept back to life. With China hot on molten salt technology, how could Oak Ridge not be as well?

A bright spot that day for Kennedy and Kutsch was that one of the original inventors of the Molten Salt Reactor had come to join the tour and hear Kennedy's talk. Dick Engel, like Kennedy, wore a bright green badge that prominently pronounced him a "visitor." Security was so tight at the lab that, unless escorted and preapproved, Engel couldn't even enter the office where he spent the majority of his career, and where he oversaw the Molten Salt Reactor Experiment as it went critical in 1964—a day many at Oak Ridge believed would change the course of America's energy future. With Engel was the sparkly-eyed and always optimistic Uri Gat, a German-born nuclear engineer who previously managed the Oak Ridge thorium program and was one of Engel's biggest supporters. Both men had retired two decades earlier, and were happy to have an invitation to come back to Oak Ridge. It didn't even matter that today's talk would be coming from some Midwestern yahoo with cowboy boots and zero scientific credentials.

So what exactly *was* Jim Kennedy's business there?

Kennedy arrived uneasily in the complex world of next-generation nuclear energy in 2009, when he found himself with what he called his "thorium problem." He was originally a little put off by Chicago engineer John Kutsch and his cast of quirky characters, for whom thorium had become a panacea for nuclear energy's dangerous aspects: proliferation and reactor explosions. While these guys lusted after thorium like conquistadors after El Dorado, Kennedy was just looking for a way to get rid of it.

Lecturing to a room of about seventy scientists and engineers, Kennedy explained how his "thorium problem" had come about. An iron mine he bought in 2001 had a significant ore deposit of a mineral called monazite. Monazite, it turned out, was full of rare earth elements that are essential to all modern technology—including American defense technology. It also contained radioactive thorium. Somehow, the United States and all the world's advanced nations had capitulated the supply and manufacture of rare earth elements to China, leaving no fully non-Chinese source for rare earth magnets, lasers, and many other high-tech products.

The rookie mine owner was now sitting on over $2 billion worth of some of the most important metals in the age of technology. Unfortunately for Kennedy, the U.S. Nuclear Regulatory Commission, responding to proliferation fears in the 1940s, classified thorium and uranium as "source materials"; that is, radioactive materials that could be made into nuclear fuel. Source materials are among the most strictly controlled substances in the world. So when the regulations began to be enforced in the 1980s, mining operations had to shelve high-value monazite and other rare earth resources to avoid costly regulatory issues. The thorium in his mine essentially made Kennedy's heavy rare earths a nonstarter economically; hence, his "thorium problem."

Kennedy and Kutsch's strategy to solve the problem—which they'd been promoting to anyone who would listen—was to get the U.S. government and industry interested in thorium nuclear energy again. With Gat and Engel in the audience, Kennedy felt like he was holding court with scientific warriors from a better, bolder America. These were the chosen few who rode the country's wave of innovation, when men dared to unlock the inner workings of an atom and launch men to the moon. In the following decades, they also witnessed a painful and calculated war against funding for science and exploration—the very things that had made the United States the economic and military superpower that it was.

Kennedy's presentation was not typical of most scientific talks at Oak Ridge. He waved his hands. He punctuated his delivery with colloquialisms. He called out bad government policies that he said buried good research deep in the Oak Ridge catacombs and overregulated a harmless rock because it contained traces of thorium. His audience wasn't jumping up and down, but his enthusiasm was still infectious. When Kennedy brought up the idea of restarting Oak Ridge's thorium molten salt reactor project, something he now believed in as feverishly as John Kutsch and the other thorium aficionados, "these two old scientists—Gat and Engel—were smiling from ear to ear. They looked like kids," he says.

After the presentation, Gehin invited Kennedy and his colleagues to a small conference room. It was a meeting that Kennedy had been lying awake at night preparing for. Here's how the scenario played out in his head: with funding from the Department of Energy, which was responsible for the budget of all the national labs, Oak Ridge would lead the commercial development of safe nuclear energy using thorium. Thorium would then be taken off the "source material" restricted list, so it could be mined and stored for future energy use. Kennedy's rare earth elements could be mined at a profit, and the United States would no longer be completely dependent on China for them. Kennedy had fantasized about it a thousand times over. His wife and best friends in St. Louis got so sick of hearing about it that they would run out of the room when he started in.

Also in the conference room was a young NASA engineer named Kirk Sorensen. Sorensen had become thorium's biggest advocate in the United States, and he was as fired up as Kennedy. He pounded the table, imploring Oak Ridge to get their nuclear research program back on track. "The conversation was, 'We're going to get this done. It's going to happen,'" says Kennedy. Almost exploding with anticipation, Kennedy held back for a moment until Gehin leisurely entered the conversation. In his Southern drawl, "Jess looks at me and says, 'So you really think you can build a commercial thorium reactor?'" The blood drained from Kennedy's face. He looked across at Uri Gat and Dick Engel. "You could see the joy drop from the faces of these two, because they were thinking that Oak Ridge was actually going to get behind this and make this happen."

Kennedy took a moment to answer. He did not believe that Gehin was being demeaning or sarcastic. His impression was that Gehin was truly asking *them* to build the next-generation nuclear reactor. However much Kennedy tried to scrub off his upbringing and make nice in a politically correct world, he couldn't.

"I lost it," he says. "I said, 'What the fuck did you just ask me?' And he asked me again, 'You guys really think you can develop this commercially?' And I just flipped out on him. 'What are you talking about? You're a *national lab*. This is what *governments* do! We're just a bunch of advocates who want to see the national labs get this done!' And he just looked me right in the eye, and he looked at all his teammates, and he goes, 'It will never happen here. Not in a million years. If you don't do it, it's never going to happen.'"

It was hard to know how much Kennedy's highly unscientific outburst rattled Gehin. Gehin was the guy running the thorium program—but the thorium program didn't exist anymore for Oak Ridge. No one should have been angrier about it than Gehin, Kennedy thought. But Gehin seemed defeated, resigned to working out his days as the leader of a brilliant team that would never get the funding it needed to create and innovate.

"People have unrealistic expectations," Gehin said in an interview a few years later. "What they don't understand is how little control of the budgets we have. This is a misconception, and it's an issue the [national] labs have. While we try to take our technical knowledge and get the DoE [Department of Energy] to move in our direction, ORNL can't say that the direction for the national lab is to use thorium. We are not the decision makers. Sometimes, people get frustrated with us."

Again and again, Kennedy and Kutsch tried to get more information about what happened when Jiang and his team of Chinese engineers came to Oak Ridge, but to no avail. "They didn't want to talk about China. They just wouldn't engage," says Kutsch. "And no matter how much you tried to engage them on geopolitical issues or on who's in the lead and who's behind, they just couldn't perceive or express risk or any lost opportunity. Nobody seemed to understand that this is a *massive* shift of geopolitics—a transfer of power. They were just like, 'Oh, the Chinese like science.' Nobody thought, How will it play out economically and geopolitically? And they would take an environmental high road when you pressed them to the wall: 'Well, this has to happen, and it's not going to happen here. So at least it's going to happen.'"

"It was a sobering, depressing experience," remembers Kennedy. A freak snowstorm buried Knoxville in a foot of snow, and he was stuck in Tennessee for two more days. Kennedy was a man who would normally spend little time wallowing in failure, but this time it was hard to shake. He got back on the plane with an intense disdain for the Department of Energy labs that would be solidified over the following years. "Our national labs, which used to be the most important drivers of technology in the world, are now museums that pretend to be national labs," he says. "All of the money that goes to them is just political money laundering. Nobody is supposed to finish the job. Everything is in neutral. The engine is shut down, and the costs keep accumulating."

Kennedy had a hard time believing that his country, throughout the twentieth century the number-one innovator in the world, would just walk away from developing the future of energy. In his quest to understand why it was so, he discovered brutal truths. He learned how “free trade” and “free market” bled the American worker nearly to death and ground the engines of industry to a screeching halt. He encountered Washington’s top leaders and Wall Street bosses who orchestrated a trade war against their own country. He watched Congress refuse to find money for science and innovation, while coming up with enough to engage in continual war. He saw solid capitalist policies he learned in graduate school buried under a pile of bogus Securities and Exchange Commission filings. He cringed as America’s greed and shortsightedness gave the People’s Republic of China carte blanche to cash in on America’s desperation for profit above country. The result was the most egregious thing Kennedy discovered: the U.S. Department of Defense was now 100 percent reliant on China for all its major defense systems. Military men and women would now go into battle hanging on the threads of a political catastrophe.

As insurmountable as the problems seemed, knowing the truth made it impossible for Jim Kennedy to walk away. With his wife running the show behind the scenes and his best friend, John Kutsch, by his side, he went looking for allies. The problems can be solved, he argued. We can bring back manufacturing and national security. We can easily win the fight for safe and clean nuclear energy. What he found were so many deaf and blind to the dire situation he believed the country was in. Others didn’t seem to care. Many threw up their hands and told him it was too hard a problem to solve. They only made him more determined.

“The United States of America went from a very obtuse understanding of the atom to building an atomic bomb in *four* years. The single most incredible human endeavor of pure science I’m aware of. The same country now says, ‘Ah, they got us. Let’s give up. Come on, everybody, get off the bench, we’re leaving the field.’”

Jim Kennedy refused to leave the field.

2

Inconvenient Ideas

To the making of these fateful decisions, the United States pledges before you—and therefore before the world—its determination to help solve the fearful atomic dilemma—to devote its entire heart and mind to find the way by which the miraculous inventiveness of man shall not be dedicated to his death, but consecrated to his life.

—DWIGHT D. EISENHOWER, "ATOMS FOR PEACE," SPEECH TO THE UNITED NATIONS, 1953

Dick Engel wasn't surprised that Jim Kennedy's pitch to revive thorium nuclear research was shut down by the Oak Ridge bosses. The world had completely changed since Engel landed in the Tennessee hills—nearly a half century before—to join a team tasked with splitting atoms into energy for peaceful purposes. From the spoils of World War II came the promise of civilian energy production; at the time, there was copious funding for research from the U.S. Atomic Energy Commission, the civilian agency charged with U.S. energy and weapons development since Truman transferred power from the military in 1947. There was also a new president in the White House who was 100 percent behind the idea of commercial nuclear energy.

Just days after he won the election in November 1952, General Dwight D. Eisenhower arrived for a secret meeting in the clubhouse of the Augusta National Golf Club in Georgia.[1] The meeting's purpose was for Roy Snapp, secretary of the U.S. Atomic Energy Commission, to brief Ike on one of the most pressing global issues in the history of mankind: atomic power.

It was certainly on his mind. Only seven years before, America orchestrated the unimaginable apocalyptic end to World War II. While there was ample reason for much of the world to welcome the end of the war, the enormous loss of life inflicted to that end haunted Eisenhower. Over 120,000 men, women, and children perished. The five-star general was one of the very few of his stature to strongly oppose bombing Hiroshima and Nagasaki in the months leading up to August 1945. In his memoir, *Mandate for Change*, Ike wrote about expressing his concerns to Truman's war secretary, Henry Stimson, a month earlier:

> The Secretary, upon giving me the news of the successful bomb test in New Mexico, and of the plan for using it, asked for my reaction, apparently expecting a vigorous assent . . .
>
> During his recitation of the relevant facts, I had been conscious of a feeling of depression and so I voiced to him my grave misgivings, first on the basis of my belief that Japan was already defeated and that dropping the bomb was completely unnecessary, and secondly because I thought that our country should avoid shocking world opinion by the use of a weapon whose employment was, I thought, no longer mandatory as a measure to save American lives. It was my belief that Japan was, at that very moment, seeking some way to surrender with a minimum loss of "face." The Secretary was deeply perturbed by my attitude.[2]

On that dreary morning seven years later, Eisenhower told Snapp that he still hadn't warmed to the massive killing power of atomic weapons. Ike was not afraid of Russia, he said; nor did he believe that such fears should rule America's foreign policy. He was, however, deeply troubled by America's growing nuclear arsenal, and told Snapp that there was certainly no need to "build enough destructive power to destroy everything."

Rather than discuss atomic weapons, Ike seemed far more interested in talking about commercial atomic energy.[3] He was well versed on the topic. Charles A. Thomas, CEO of the Monsanto Chemical Company and a big player in the Manhattan Project, put a bug in Ike's ear, pushing for industry control of both commercial and military nuclear reactors. Snapp explained that a lot of work was already under way in the private sector. General Electric, Union Carbide, and Westinghouse were in the midst of developing reactors. But under the Atomic Energy Act of 1946, the government was still required to maintain ownership of all nuclear facilities and fissionable material used to fuel reactors.

Six months later, on April 29, 1953, in a speech to the United Nations, President Eisenhower announced a national policy to develop atomic power for civilian use.[4] He met little opposition. It was a time when the United States hungered for discovery and innovation. Patriotism fueled public support for exploring the far reaches of the universe, down to the center of the atom. The call came from the highest office in the nation and percolated through universities and public agencies, and all the way to rural elementary schools. It was this national fervor that brought Dick Engel to Oak Ridge National Lab upon his graduation from the University of Toledo in 1953. There he would work under one of the fathers of nuclear energy, Alvin Weinberg.

Weinberg came to Oak Ridge in 1945 after his work on the atomic reactor at the University of Chicago.[5] Like several of his fellow Manhattan Project scientists, Weinberg was never sold on the idea of actually using the bomb, and he was happy to refocus his research on new reactor designs for commercial energy. Weinberg, who would become the lab's director, also knew that he needed an army of talented scientists and engineers to take nuclear reactors to the next level. Together with Captain Hyman G. Rickover (later Admiral Rickover), who was leading the effort to develop nuclear power for the U.S. Navy's submarine fleet, Weinberg created the Oak Ridge School of Reactor Technology. The program became the premier school for nuclear engineering in the world, and at twenty-two, with a bachelor's degree in chemical engineering, Dick Engel knew it was exactly where he wanted to be. The native Ohioan received his letter of acceptance in 1953 and headed for the hills of Tennessee to ride the wave of an emerging revolution under the biggest names in field.

Engel spent his first night in Oak Ridge at the Guest House, a classic two-story hotel built during the Manhattan Project days.[6] A decade before him, dignitaries like J. Robert Oppenheimer, Enrico Fermi, and secretary of war Henry Stimson came to sleep in the cozy guest suites and enjoy the fancy hotel restaurant. For Engel, the luxury didn't last. The next night, he was in the cramped dorms with all of the other students. "If you were single and didn't have any special connections, you were assigned a room in a dormitory which was about ten feet long, maybe eight feet wide."[7] The shower and bathroom facilities were in the middle of the building. There was also a communal kitchen that Engel never used, because he found the town of Oak Ridge to be full of good restaurants.

The University of Toledo didn't prepare Engel for the intensity of the nuclear program at Oak Ridge. "I never worked so hard as a student in all my life, before or since," he said. There were students from all over the country, all reeling from the pressure. Engel made friends with a student from the academic powerhouse Caltech, another with a master's degree in physics from the University of Chicago. There were government and industry employees who were already established nuclear engineers. "So I, with my piddling B.S. degree, was sort of at the low end of the scale. But everybody agreed that it was a more intensive education program than they had faced before."

After Engel finished the yearlong program, Oak Ridge offered him a full-time job. In his heart, he was a chemist more than an engineer, which helped him make up his mind: "I decided that the work I would be doing there was more chemically oriented than the kind of work I would have been doing for Westinghouse or GE." Both General Electric and Westinghouse were working on nuclear reactors for submarines, but staying at Oak Ridge and developing new reactor designs for different applications had more appeal to Engel. The one condition was that he had to get out of that dingy dorm room. Engel told them he'd take the job if they got him an apartment. "And the next day, I had a one-bedroom brick apartment for $70 a month—all utilities paid."

At the time, Engel was one of thousands of single people filling the excellent government jobs in the "Secret City." A lot of the support staff were single women. Most of the engineers were single men. Still, Engel found it hard to have much of a social life. "I am not an extremely social person," he says. "I'm sort of private and quiet." Fortunately for Engel, a local Methodist church played matchmaker for some of the shy, awkward engineers by holding socials and get-togethers. Here, the very reserved Engel was scooped up by the extremely outgoing Lorraine, who worked in one of the lab's many administration offices. Engel married Lorraine and grudgingly adapted to her social schedule when necessary.

At the time, Engel's big boss, Alvin Weinberg, oversaw several reactor projects. One of the standouts was the Aircraft Reactor Experiment.[8] The project's goal was to make a nuclear reactor small enough and powerful enough to drive an airplane. Knowing that the U.S. Navy was all-systems-go with nuclear-powered submarines that could stay underwater for several months, air force brass were determined to have nuclear-powered airplanes that could stay in the air for a month at a time. The challenges were enormous: the actual reactor needed to be small enough to fit inside a plane, but the radiation shield needed to weigh 50 to 100 tons, and the reactor would have to operate at 870 degrees Celsius. If it crashed, contamination would spread across thousands of square kilometers.

"What caused otherwise sober, even brilliant, engineers to conclude that nuclear flight was not crazy?" wrote Weinberg in his autobiography, *The First Nuclear Era: The Life and Times of a Technological Fixer*. "In part, it was fear of the Soviet Union and the difficulties in developing ICBMs (intercontinental ballistic missiles). In part, I suppose it was autocatalytic optimism. These were bright people, and they naturally came up with new ideas." Weinberg really never believed the project would get off the ground, and in 1959 he confessed his feelings to Herbert York, director of defense research and engineering at the Department of Defense. Already $880 million had been spent on the project, and one of the first things that John F. Kennedy did when he came to office was to scratch it from the Atomic Energy Commission budget. Later, two AEC directors cornered Weinberg and accused him of treachery and dishonesty. He wasn't used to making enemies and wasn't sure what it meant for his future, but he feared that his candor with York would cost him dearly well into the future.

Although a nuclear-powered airplane never flew, Weinberg was proud of the many discoveries that came out of the project—most importantly, the idea of liquid nuclear fuel, a much different model than the solid rods of uranium fuel that had been the only game in reactor technology up to that time. A chemical engineer named Ray C. Briant who joined the Oak Ridge team in the middle of the Airplane Reactor Experiment thought a nuclear-powered airplane was a preposterous idea. In a submarine, constant water circulation can keep the reactor cool, and weight is not an issue. In a land-based reactor, highly pressurized water in a massive contraption cools the reactor and generates steam for the propulsion system. But in a plane flying at high altitudes in a thin atmosphere, all of these methods go out the window. According to Weinberg, Briant ridiculed the operation, saying that at 870 degrees Celsius, with no way to cool them down, the solid uranium rods would come out looking like spaghetti. Briant was a chemist, and what he imagined that the physicists and engineers couldn't was how certain liquid substances behave at certain temperatures. If they could put actual uranium fuel into something liquid, it would not deform or break apart under high temperatures. But what kind of liquid would work?

Two other researchers, Vince Calkins and Ed Bettis, found the Holy Grail: molten salt.[9] For the MSRE team, the salt had to have certain special properties. What turned out to be the best bet was a salt mixture of fluoride bonded to lithium and beryllium, with lesser amounts of other metals. The liquid fluoride salts worked well in a mechanical system and were excellent at dissolving radioactive fuel. They were also perfect for handling the temperatures experienced in a nuclear reactor. The system could operate as low as 450 degrees Celsius (the melting temperature of the salt), and the fuel was stable even above 1600 degrees Celsius.

The molten salt technology was deemed brilliant and quickly green-lit by the Atomic Energy Commission as a new project called the Molten Salt Reactor Experiment (MSRE). The MSRE became known as the "chemist's reactor," because it was the brainchild of chemists and a product of liquid fuel chemistry rather than merely the physics of fusion. Its inventors believed that it would be a safe way to make a civilian nuclear power plant without pressurized steam as a coolant, or the threat of a runaway reactor—a serious concern even twenty years before the Chernobyl disaster.

When Engel transferred from another project to work on the MSRE, the chemical engineer felt right at home. The project was already under way, and the engineers were well along with the design of the reactor's core—the place where they put the fuel and where the heat is generated. As construction moved forward, Engel moved his office to the reactor site, where he helped plan the test program and train the operators. Engel worked the calculations and processed data mainly with his go-to slide rule and a pencil, but he had state-of-the-art technology too, such as an early calculator similar to devices used by bank tellers at the time: "You could multiply with it, and if you were really clever, you could make it do square roots," he says. There was also a very large, centrally located IBM computer for very complex reactor computations. After inserting long trays of punch cards, researchers hoped to get their results a few hours later. "The MSRE was the first reactor that I know of that had an online computer installed, and it filled the space about the size of a room," Engel says. "Sometimes, you put one card in wrong, and nothing went through." At the time, one of the most pressing questions for Engel and his colleagues was whether the salts would freeze up and "put us out of business." Salt that froze solid would be a disaster for the system, particularly in the smaller pumps. The salt mixtures with the right chemistry for a reactor had high melting temperatures, and could turn solid if the temperature dropped below 450 degrees Celsius.

Overall, the molten salt reactor was not an extremely complicated design. Alvin Weinberg was famous for calling it "a pot, a pipe, and a pump," a phrase originally coined by Eugene Wigner, a fellow Manhattan Project physicist. Paul Haubenreich, the engineer in charge of the project, agreed. "The operation of the MSRE was not too difficult—wasn't too burdened with problems," he said in a 2012 interview.[10] Haubenreich was a native Tennessean who, unlike Weinberg and Engel, spoke the same language as many of the technicians hired from the neighboring East Tennessee countryside. "One of the things they all had besides hound dogs under the porch were old cars out in the yard that didn't run very well. And every one of these country boys had to learn how to raise the hood and fix things." Once, when a cable for lowering sample capsules into the fuel tank was accidentally cut, Haubenreich was crushed. "I said, 'Oh my goodness, we're out of business.'"

One tech told the boss he was positive they could fix it.

"I said, 'How are you going to do it?'

"'I don't know, but we will fix it!'"

And they did. Another time, a skunk went to sleep in the trashcan, and the techs got it out without the slightest stink.

One thing that Engel didn't have to concern himself with was the cost of the program. "We never gave a thought to what the budget was. That was 'them folks up there' worrying about stuff like that. You had your job, and you were fully absorbed in what you were doing." Only high-level management like Alvin Weinberg worried about how the budget was distributed among experimental work, reactor construction, and other essential components. Another factor that didn't worry the team was whether the MSRE would be successful. In those days, Engel says, "You didn't have to guarantee success, because if that failed, you could try something else. In the current climate, you don't even start until you can pretty much guarantee that you're going to be successful."

In 1964, Haubenreich's team first started putting liquid salt into the MSRE to check out the pumps and valves and see if the electrical system worked. Then they included depleted uranium (not fuel grade) and circulated it to make sure the chemistry behaved predictably. "Now we're ready for criticality," says Engel. After years of painstaking work, overcoming glitch after glitch with brainpower and slide rules, Engel slept peacefully at home the night in 1965 the reactor went critical. For another four years, the Molten Salt Reactor Experiment was a dream machine, performing almost exactly to design. It became a beloved member of the Oak Ridge family, marked with a ventilation stack of superheated air from the hot tubes underground. "In the wintertime, when it was bitter cold, all the birds in Anderson County perched at the top of that stack to get warm," remembers Haubenreich.

"It worked out very well," says Engel. "Valves got plugged with oil vapors and things of that sort—things that had to be fixed—but there were no showstoppers in the system."

Haubenreich and Engel published a detailed account of their work, "Experience with the Molten-Salt Reactor Experiment" in *Nuclear Applications and Technology*, in 1969.[11] In their conclusions, they wrote, "From the months of operation and experiments, a very favorable picture emerged. In properly designed equipment, handling the high-melting salt proved to be easy. Maintenance of the radioactive systems was not easy, but there were no unforeseen difficulties, and control of contamination was, if anything, less difficult than expected. When measured against the yardstick of other reactors in a comparable stage of development, it is seen to be indeed remarkable."

Alvin Weinberg was thrilled. "After such a success, our confidence soared, and we expected that the AEC [Atomic Energy Commission] would support a much larger molten-salt [thorium] thermal breeder [reactor]," he wrote.[12] The plan was to build reactors that could generate 1,000 megawatts of electricity per plant—enough to power a million homes. Also, because the United States was concerned that uranium was in short supply, they configured the reactors to "breed"—they would make more fuel than they consumed. (At the time, the country was worried about running out of uranium, a needless concern, as it turned out; there are plentiful uranium sources in the States.)

The molten salt reactor was tremendously safe. Unlike the GE model, or those built by Westinghouse and others using highly pressurized steam or boiling water as a coolant, there was no fear of the containment structure exploding. Twenty years before Chernobyl and more than four decades before the Fukushima disaster, Alvin Weinberg had a working reactor in his pocket that the Oak Ridge team believed would, over time, become the future of safe, clean nuclear energy. America could one day have unlimited energy made from the splitting of an atom.

After four years of successfully generating power and tantalizing the nuclear community around the world, Oak Ridge shut down the Molten Salt Reactor Experiment in 1969. Weinberg and the others were anxious to work on the next-generation reactor, but it was not to be. In 1972, the Atomic Energy Commission unceremoniously pulled the plug on any new molten salt research.

"Why didn't the molten-salt system, so elegant and so well thought-out, prevail?" wrote Weinberg:

> One, politically, all of the reactors that were operating were solid fuel and water cooled, so the technology was unknown to many in the field. Perhaps the moral to be drawn is that a technology that differs too much from an existing technology has not one hurdle to overcome—to demonstrate its feasibility—but another even greater one—to convince influential individuals at organizations who are intellectually and

emotionally attached to a different technology that they should adopt the new path. This, the molten salt system could not do. It was a successful technology that was dropped because it was too different from the main lines of reactor development. But if weaknesses in other systems are eventually revealed, I hope that in a second nuclear era, the molten salt technology will be resurrected.

Dick Engel didn't know who the decision makers in Washington were, let alone their reasons. What he did know was that he was out of a job, and his prospects did not look good. "Nobody else was much interested in liquid fuel systems," he says. Any new nuclear development was considered unnecessary and, thus, out of the question, as far as the U.S. government was concerned.

When it was over, "I felt let down," says Haubenreich,

> But I was aware that times had changed. Weinberg was not a convenient but an inconvenient originator of ideas. During the 1950s and 60s the climate was such that the AEC gave him a couple of million dollars and said, "See what you can do with that idea," and that was a lot of fun. Weinberg ran with that ball and thought of another one or two while that was happening. But in the early 70s, there wasn't any place for a person of his stature. Even though he was extremely well qualified to guide nuclear programs, the AEC and Washington felt they had plenty of people up there who could guide the programs . . . *and we want people somewhere down in the laboratories that know what we want done and not to bother us with all these ideas.*

Weinberg became very outspoken about his concerns with the safety of light-water reactors, and was soon a thorn in the side of California congressman Chet Holifield. Holifield, dubbed "the Atomic King," was on the Joint Committee for Atomic Energy and was so fervent about commercial nukes for the world that he proposed building enormous reactor parks in San Onofre, California. He had deep ties to the light-water reactor industry, and labeled anyone with any objection to his ideas "kooks." Holifield approached Weinberg in 1974, saying, "Alvin, if you are concerned about the safety of reactors, then I think it may be time for you to leave nuclear energy."

Weinberg was blown away. "It was apparent to me that my style, my attitudes, and my perceptions of the future were no longer in tune with the powers within the AEC," he wrote. In late 1972, he was fired.

Being cut off from government funding could not stop Alvin Weinberg's mind from churning up new ideas or pushing those he fervently believed in. He promoted the thorium molten salt reactor to any who would listen. He became deeply concerned about the effect of fossil fuel consumption on the atmosphere. He started his own think tank, the Institute for Energy Analysis, which became the premier research institution on carbon dioxide buildup. "Continued energy demands during the first few decades of the next century will push atmospheric carbon dioxide concentrations to levels which warrant serious concern, even for the low energy growth case," he wrote in a 1974 article in *Science*. "I went from office to office in Washington, curves of the carbon dioxide buildup in hand," Weinberg wrote in his autobiography. "I reminded them that nuclear energy was on the verge of dying. Something must be done. I almost screamed."[13]

In 1974 Weinberg got a front-row view of how things worked in Washington when he was appointed director of the Office of Energy Research and Development. Capitol Hill couldn't have been a worse place for him to land, and he'd never felt farther from the hills of Tennessee. He had a beautiful wood-paneled office in the Old Executive Office Building and a small American flag on his desk—so horribly different from his laboratory office back at Oak Ridge. Depression enveloped him—partly because he still felt the loss of his wife, Margaret, five years before, and "partly confusion in my own mind as to what in the hell my Office of Energy Research and Development was supposed to do," he wrote. "But in Washington, one must always appear to be busy and to be creating 'policy' even if there is little to keep one busy, and even when no one has a clear idea as to where policy ought to take one."

Weinberg was so miserable that he began to take Valium every day, and couldn't escape thoughts of suicide. He found himself half wishing he'd be mugged on his walk back to his Watergate apartment after dinner at the Cosmos Club, and it would all be over. "This depression stayed with me through 347 days in Washington," Weinberg wrote. It was one of the worst years of his life.

There was a brief glimmer of hope for new nuclear research in 1976, when Washington became concerned about bomb-grade fuel getting into enemy hands. The catchphrase became "nuclear proliferation." Classic light-water reactors produce plutonium when they burn fuel, and they need enriched uranium to get up and running. So the Department of Energy reached out to its national labs to develop a new reactor design that would be safer than the status quo. After spending his time tying off the loose ends of the defunct molten salt program, Dick Engel was thrilled to be again actively developing a new reactor based on molten salt chemistry. The Oak Ridge team knew that if they added thorium to the fuel, their reactor would not produce nearly as much plutonium at the end of the cycle. It would not explode or become a proliferation threat. It would be the ultimate safe reactor. Unfortunately, the program ended soon after it began. "One of the conclusions was that the molten salt reactor may be viable, but it was not sufficiently developed," says Engel. "It didn't make the grade, is what it comes down to."

For decades, Alvin Weinberg kept waiting for another nuclear renaissance, but it was not to be. Not only was the nuclear establishment against him but the growing environmental movement battled new nuclear as well. This infuriated Weinberg to no end, as his work was deeply based on his concern for the environment. "Since nuclear reactors emit almost no carbon dioxide, how can one be against nuclear energy if one is concerned about carbon dioxide?" he would argue. "To my utter dismay, indeed disgust, this is exactly the position of some of the environmentalists."

For forty years, light-water reactors were the only game in town. No new designs were built. Chernobyl blew up. Then Fukushima happened. The fear of nuclear energy in the minds of people and governments grew.

Alvin Weinberg died on October 18, 2006, in Oak Ridge, Tennessee, at the age of ninety-three. Had he lived four more years, he might have been delighted by a renewed interest in his molten salt reactor by a small but eager group of engineers and advocates. It might have seemed strange to him that the people digging through yellowed handwritten notes and dusty files at Oak Ridge were not renowned physicists or industry engineers. They didn't have Weinberg's credentials, his background, or his long history in the field of nuclear engineering. They did, however, share his vision of a clean energy future for the United States and the world.

What may have surprised Weinberg even more was that Jim Kennedy, John Kutsch, and NASA engineer Kirk Sorensen weren't the only ones interested in those discoveries made at Oak Ridge so long ago. Another person came to the Tennessee Hills from very far away. This man had strong scientific credentials and deep pockets full of government funding for nuclear science and research. He came looking for clues in his quest for an energy panacea that would fulfill the exploding energy needs of his 1.3 billion countrymen. He looked for a method that didn't need water for cooling, because his country had vast expanses of geography without wide rivers or coastlines. He searched for a way to burn thorium, not uranium, because his country had nearly infinite thorium reserves. And because this particular scientist was also a very smart businessman, the technology must become a bona fide commercial success. The man's name was Dr. Jiang Mianheng.

3

The Misfit

In 1967, the year that the Oak Ridge Molten Salt Reactor went critical, four-year-old Jim Kennedy watched his green Missouri hills disappear from the window of a giant jet airplane and transform hours later into swaying palm trees and shimmering blue ocean as far as the eye could see. The move to Southern California was a grand adventure for the Kennedy clan: six rowdy siblings under eight years old, their beautiful mother Elizabeth pregnant with number seven, their bright, charismatic father leading the charge. The transfer across the country to work at Monsanto's plastics division in Orange County, California, was a stepping-stone in the senior Kennedy's career. Unlike his contemporaries, Mr. Kennedy worked a sixty-hour week, unheard of at the time. From a young age, Gerald Kennedy had a deep drive to succeed. He pushed himself physically and mentally to an extreme—raising pigs and funding his way through college by investing his profits in the stock market. As an Olympic-level boxer, he punished his body and his opponents. While he hadn't learned interpersonal skills growing up in his deeply dysfunctional home, he was charming and thoughtful to colleagues and subordinates at work.

He could not have been more different at home. "My dad would come home from work and sweep his finger over a doorjamb. If there was any dust, he would clear the nearby shelves with a sweep of his arm," sending everything crashing to the floor while the children watched, frozen. "You can see it in the old photos of my mom and us kids. We all had to look perfect and be perfect," Jim remembers. All of the domestic issues for the family were left to Elizabeth. "The pressure was on our mom."

The family of eight squeezed into one of the thousands of unremarkable postwar ranch homes on a small lot in the exploding suburban landscape of Southern California. For little Jimmy, California was paradise. On weekdays, the three-block walk to La Veta Elementary for kindergarten was a grand journey—traversing the neighbor's retaining wall to cut his travel time, sucking on honeysuckle blossoms and terrorizing ant colonies along the way. In the summer, he and his older brothers went door-to-door selling avocados from their backyard trees and vegetables from their father's garden. Summer nights, neighborhood kids played kickball and flashlight tag long after the streetlights went on. "It was a time when we were like everyone else," Jim remembers. "We had friends. We fit in." All of Jim's aunts and uncles called the little rambler "James West" after Robert Conrad's character in the popular 1968 television series *The Wild Wild West*, because he never, ever, took off his most prized possession: his black leather cowboy boots with white and red diamonds. "My only ambition in life was to become a cowboy," he remembers.

Two years after he landed in California, six-year-old Jim Kennedy's childhood shattered in an instant when a blood vessel ceased to deliver oxygen to his mother's brain, terminating her bubbly vivaciousness with a massive stroke. For nine months, Elizabeth Kennedy lay in a coma in an Orange County hospital. The children, shell-shocked, were first displaced to aunts and friends, but eventually trickled back to the cramped house in the City of Orange. Elizabeth remained in ICU, and Mr. Kennedy hired a series of nannies—"Really, just inexperienced babysitters, drifters or extended family members who stayed with us while they worked out problems of their own," Jim says. Sometimes the chaos had its perks. Sunshine, one of the live-ins, had a boyfriend who was in the Hells Angels motorcycle gang and rode a great big Harley-Davidson with a high-backed seat. "Sunshine liked me, so she would have her boyfriend give me a ride to school on the back of his chopper when I was seven." The kids ate chicken potpies and TV dinners every night. "I loved it."

Finding the long weekends as a single parent of seven intolerable while Elizabeth entered a convalescent hospital to begin a year-long period of rehabilitation, Mr. Kennedy bought his two oldest boys, Jerry Jr. and Joe, brand new Huffy ten-speed bikes. He bought himself a twelve-speed Peugeot. Athletic and very fit in his mid-thirties, Mr. Kennedy's plan was to spend his weekends riding down the paved Santa Ana Riverbed all the way to Newport Beach, a thirty-five-mile round trip. When six-year-old Jim saw that there was no bike for him, he was not deterred. As luck would have it, the neighbors next door tossed out a worn girl's bike. *Perfect.* He hopped on. His short little legs barely reached the pedals as he stretched his toes as far as he could. For the next two years, if you were ever walking or biking the Santa Ana Riverbed, any weekend, you could have seen a little boy with sandy blond hair, small for his age, on a girl's bike way too big. The boy's high-octane legs spun in a furious whirl to keep up with his father and two older brothers, who did everything they could to prove that little boy wasn't worthy of being on the trail with them. They never could.

The evenings were more somber. Mr. Kennedy would yell for Joe and Jerry Jr. to come to his bedroom for a competition. Which one of his two oldest sons could recite the best prayer? Uninvited, little Jimmy would be right there with his older brothers, pouring his heart into every word as he knelt and prayed. His father would often choose Jim's prayer as the winner, and he would earn the honor of sleeping in that great big bed. Spending the night next to his father was exciting, but falling asleep was difficult, his little mind consumed with thoughts of God and his mother. God could hear his voice, he was sure. If he prayed hard enough, he knew that he could move God's hand. And therefore, he held the life of his mother—the one person who he felt truly loved him—in his small hands.

Finally, two years after his mother's stroke, little Jimmy's prayers were answered. Elizabeth came home. But the Elizabeth Kennedy who arrived in a wheelchair did not look like his mother. The mother who left him was bubbly and fun, driving a 1968 Barracuda and flirting her way out of a speeding ticket by offering to let the cop drive her sporty car. This mother could not walk. She was mostly blind. She seemed to be staring into a far-off place in her mind. Jim was crushed. There was no hug, no "I missed you." For the first few months, his mother didn't even know his name. She was weak, confused, and unable to care for herself—as helpless as a baby. Every day, eight-year-old Jimmy would stare at his mother, trying to understand what God had done to her and why.

Elizabeth slowly recovered basic motor skills with the help of Jim and his two older brothers. They took turns encouraging her to take a few steps or work her arms and legs on a pulley system that ten-year-old Jerry Jr.

constructed on his own initiative. It was during this time that Jim began to feel a deep connection to his disabled mother. Toward his father, he had no such bond: "My father felt that he only had two children; Jerry Jr. and Joe, the two oldest." By being third, he could have chosen to be largely ignored, to keep a low profile to escape the severe discipline that the older two received. "He didn't ask for anyone else. He didn't look into anyone else. But I *refused* not to be acknowledged. So he would yell out, 'JERRY . . . JOE, get down here now!' and I would come running, determined to be part of whatever it may be—good or bad." The younger children remained mostly invisible to Mr. Kennedy. "But I demanded to be recognized, even though it came with a price."

Mr. Kennedy, who had amassed significant savings by investing in the stock market before Elizabeth's stroke, was now essentially bankrupt. Medical bills and rehabilitation had burned through all of the family savings. There was no longer money to hire live-in help. Elizabeth became stronger, but she was still severely handicapped, and keeping control of the household was utterly impossible. Neighbors occasionally dropped off home-cooked meals for the family, what Jim's brother Joe remembers as a great experience during his childhood. The Kennedy kids ran wild. "We pilfered the neighbors' trash on trash days," remembers Jim. "Someone's discarded toy would become a most prized possession. The neighbors started calling us 'the Dirty Kennedys,' and their kids shot BB guns at our house and at us." In 1970, disabled, struggling to care for her own basic needs, Liz Kennedy became pregnant for the eighth time. Spurred by the financial stress of Elizabeth's illness, ever greater work demands, and another child in a shrinking house, Mr. Kennedy became increasingly volatile and violent.

"If you did something wrong, you could expect a 'spanking,' which, under the best circumstances, required you to grab your ankles while he swung his leather belt at your backside. In a fit of rage, it could be a fist or a flying object." The family was extremely religious, what Jim remembers as an almost medieval form of Catholicism. "We believed that corporal punishment, dispensed randomly, was a normal part of life and surely preferential to God's punishment later." In Catechism class, little Jimmy would try to make God happy by positioning himself so that his small knees landed on the brass heating grate. When the cold metal dug deep into his kneecaps and the pain was too much for him to bear, he stayed longer and pushed harder. *Pain is sacrifice*, the small boy told himself. *God likes sacrifice.* "It was at Catechism that I first learned that other kids were not 'spanked' by their parents," Jim says. "I felt sorry for these kids, knowing they would grow up weak." The word *love* in the Kennedy house was deployed only after a disciplinary session got out of control. As Mr. Kennedy's rage waned, "My father would tell you, 'Come here. Sit on my lap, and tell me you love me,' which you wouldn't say until he put you in a headlock and he was literally choking the life out of you. Love was never given willingly: it was extracted by force."

In late 1975, when Jim was twelve, his father moved his family back to Liz's hometown of St. Louis after a two-year stint in Connecticut. Each morning, Mr. Kennedy would go to Mass at 6:00 A.M. and then to the office. The Kennedy children all attended the church's parochial school. Mr. Kennedy's frustrations began to erupt more frequently, commonly intertwining religion and punishment. "If you were not participating in the evening rosary with sufficient reverence, you could become subject to a slap on the head or a full-blown spanking in any of its manifestations." There was no such thing as sibling camaraderie. Instead of the kids taking some solace in one another, each became the others' saboteur. "My father was a broken person, and he would actually play his kids off each other. We grew to distrust each other. We would form temporary and shifting alliances. We were never supervised, so our aggressive behavior became just another form of communication." By adulthood, every Kennedy sibling had one or more of their front teeth missing due to an altercation with another sibling.

Relocation to St. Louis landed the family in a quiet upper-middle class neighborhood. Loud and out of control, the Kennedys made a dramatic entrance: "We quickly assumed the honor of the most despised family in the neighborhood." Jerry Jr., the oldest, was into fast cars and terrorizing the neighbors. One of his most infamous stunts was hanging a noose from the elderly neighbors' tree. In the noose was a piece of cardboard with the name of the couple's dog written on it. Joe, the second oldest, was into wild parties with drugs and alcohol. Jim took after them both. Neighborhood kids flocked to the house. "We'd all go over to the Kennedys' because surely *something* was going to happen," remembers Mark Haselhorst, Jim's best friend from high school. Someone would light a lawn mower on fire. A motorcycle was in the bottom of the pool. Jerry Jr. would jump his van over a dirt berm in the backyard. There was always beer. Someone would shoot a rifle out the second-story window. "I can tell you from an outside perspective, it was the worst and most dysfunctional home that anyone could have ever imagined," says Mark. "I still have vivid memories of exactly what that house looked like . . . the odors . . . the decay." He half-jokingly compares the Kennedy home to the Addams Family house, and remembers the first time he saw it—the one house in an upscale neighborhood with gutters falling off and a broken car sitting on blocks in the driveway.

Mark remembers tragic scenes that, to a boy from a relatively functional family across town in a blue-collar neighborhood, he couldn't wrap his teenage head around. "I remember one day going over, and Jim's mom is sitting in the kitchen sink, yelling 'Help! Help!' because she's physically disabled, and she slipped trying to clean it, and had probably been there for hours. It was just horrific shit like that. As a kid, you don't know the impact of that. But as an adult you look back and you wonder how many hours this woman sat in the sink unable to lift herself out in this disgusting kitchen, or how or why she got up there in the first place." There were plumbing problems. Toilets did not always flush. Faucets leaked. Eventually the floor joists rotted out. "This was once a nice home in an upper-middle-class neighborhood, and we basically destroyed it," Jim says.

What his sister Mary remembers of Jim Kennedy during that period was a dark figure, engrossed in the game of the era that took hold of millions of social misfits: Dungeons & Dragons. "I felt like Jim was isolated from the family," she says. "I think he was always trying to escape, to be invisible. He was always hiding in the neighbor's barn or the fields or the woods. He was just gone." In the forested hills behind the Kennedy home, in the richest neighborhood in St. Louis, a twenty-five-year-old orphan named Pete, living off a multimillion-dollar trust fund, befriended Jim and Mark. In Pete's mansion, the two high school boys got a taste of how the children of Missouri's überwealthy lived—with raging parties exploding with drugs, orgies, and alcohol. Mary remembers Jim hoping that Pete would adopt him. "I don't think Jim wanted to live with us, because our family was too chaotic."

Unfortunately for Jim and his siblings, as Mr. Kennedy became more successful at Monsanto, he became more violent at home. The Irishman—who trained as an Olympic boxer—"was a Scud missile," remembers Mark. "If you were in his path, God help you, he would unleash on you." Mark remembers sitting outside the window with other friends in the dark of night and watching Mr. Kennedy scream at his children to get dressed for church. To Mark, one of the Kennedy children was particularly singled out. "I think Mr. Kennedy hated Jim with a passion. I think he disliked the other kids, but there was something about Jim. He would see Jim and go psycho. You've got to understand that Jim was woken up by a father who drove a very nice car, was successful at his job, lived in a high-end neighborhood, and who chose to wake up his son in the morning by punching him in the head. That's how Jim started his day." Jim says that he barely recollects the abuse he received at this stage in his life except that it was so frequent and so severe. To avoid being punched while still in bed, he and his brothers started sleeping in closets and attic spaces.

By the time that Jim was a teenager, he discovered that his father had an Achilles' heel. "Spankings" with his father's belt were commonplace for Jim and his brothers. "He would pull off his belt and say, 'Bend over and grab your ankles.' My brothers would jump, cower, and cry before they even got hit. Not me. When he would hit me, I would look back and smile. And that would make him mad, and he would hit me harder. I would just keep smiling. He would get so angry. He would lose his mind. And no matter what, I would keep a smile on my face and maintain eye contact." The older Kennedy would have to stop and sit down and hold his heart until he composed himself enough to continue the beating. "After about four to five times of that, he never 'spanked' me again." While the corporal wakeup calls—a punch to the face, and sometimes furniture or flatware used as projectiles—continued, Mr. Kennedy never used his belt on Jim again, even though he continued to spank the other children for years. It was the first significant victory in his life.

"Back then I hated my dad for *what* he was, not for hitting us. He managed the house like a concentration camp. He played us against each other. Our strengths were used against us. Every single one of the kids was exceptionally athletic. If you were really good at soccer and wanted to play on a team, he would use that as leverage against you. He would not let you go to the games or practice unless you did extra chores. Not just extra chores, but basically everything he could think of. No one else would be tasked with extra chores, just the person he could leverage against. Eventually it was better to drop all outside interests and be left alone." Because of this tactic, every one of his siblings learned to be underachievers. *Have no interest whatsoever, and he has no leverage on you.* "The physical abuse left cuts and bruises, but the psychological abuse left deep scars that grew into lifetime disabilities for most of my siblings."

Jim had one dependable ally at the home at the end of the cul-de-sac on Reauville Drive: his mother. Jim frequently skipped school, and on Elizabeth Kennedy's good days, they would stay home together and share jokes. He'd ask her to tell him stories about her childhood, seeking a deeper connection and truly enjoying himself. She still called him Jimmy, and he knew he was her favorite. "My mom and I always had a great relationship. I suspect that none of the other kids had this type of connection." The connection fueled him: "I could feel her inner strength." He also loved his mom's sense of humor and her resilience. He would frequently chauffeur her on errands. On one of their outings, he failed to completely close the passenger-side door after placing his mom in the front seat. He also failed to slow down when he took a sharp turn. The combination of his car's vinyl seat and his mother's polyester pants was a recipe for disaster. Liz Kennedy became a flying projectile, landing hard on the asphalt on her backside and taking on some minor lacerations. Once they both knew that Liz was okay, they burst out laughing. "My mom was pretty fucking tough," he says. Even severely disabled, Liz always tried to do laundry or to cook. "Standards were low and there was the occasional disaster, but she did manage to put food on the table in the early years," Jim remembers. She would cook all day, and dinner would be devoured by the eight children once it hit the table. Even in her severely handicapped state, she would occasionally go after her unruly brood with a broom. "You have to fight to be a good Christian," she said, laughing. Although there was no way she could protect the children from Mr. Kennedy's relentless abuse, on occasion she would stand up to her husband. "I remember she would just stare him down. It was impressive because she was a completely broken person. It took incredible courage, because she knew what he was capable of."

By the time he was in high school, Jim walked from his house on the farthest boundary line in the school district to Kirkwood High, more than two miles away. There was a bus, but he never rode it once. "He was the only kid that I knew who, at the age of fifteen, would deal with problems by taking long walks as though he was an adult," remembers Mark. By the end of his freshman year, Jim had a car that ran sporadically, but he still loved the long walks to and from school. And during those walks, he began a lifelong process of dividing the world into that which was bad and wrong and that which was good and right. It was where he honed an internal moral compass that would guide him throughout his life. Good: *Mom; the woods; Mark's normal family, and parents of other friends who sort of adopted me.* Bad: *my father's temper; my home; the way my siblings interact with each other.*

Jim's broken home life made him a complete oddity at Kirkwood High, a massive school with a student body of more than 2,000. "He basically formed his own personality, and that's where he got his nickname, *Freak*," remembers Mark. "Because in high school, if you don't fit into a certain clique, what the hell is wrong with you? You're not a jock. You're not a burnout. You're not a band geek. You're a freak." What Mark (who Jim would later describe as his only "normal" friend from those years) found to be most enigmatic about Jim was that most kids who are weird or different or slight of build would generally get the shit beat out of them. "He's physically small. He had really long curly hair. He tended to dress in black. He was a prime candidate for being bullied, because he's a misfit. But you would never see him shoved against the locker or his pants being pulled down or being abused in any way. I think he made the other kids uncomfortable." Jim tends to agree. "I walked down the hall and people got out of my way. I just didn't care about anything. That's why people in school were afraid of me. That's why people didn't want to fight me. Because I didn't really care what happened." The whole time they were in high school, Mark only saw Jim in one physical altercation. It was in a mall when another kid took a swing at Jim, and Jim punched him back.

But signs were beginning to emerge that beneath the hard exterior were elements of charm. "A lot of my friends' parents sort of adopted me, and felt bad for me, even though I had long hair, rarely wore shoes, and looked dangerous." At five foot six and 120 pounds, he was by no means physically intimidating. Among his stoner friends, Jim was always the one nominated to stand in front of the liquor store and ask adults to buy beer. "I was never afraid to approach anyone." He rarely got turned down. Once, he remembers an off-duty cop even apologized to him. "He was like, 'Sorry, I really can't. But I sure hope you get some.'" Jim Kennedy spent every day of his life—using every skill he could muster—to get what he needed in a house where all the cards were stacked against him. Negotiating in the outside world, he found, was far simpler, even if he had no clue what the rules were.

It wasn't only adults that saw something attractive about the Freak. Much to Mark's chagrin, his buddy Jim had what Mark considered to be the most coveted quality any teenage boy could have: he was a chick magnet. What was so distressing to Mark was that Jim completely failed to put it to use. "Jim was very cute. He looked like Tom Cruise. Girls sought him, and he didn't give a fuck." In their freshman year, Mark's girlfriend, a beautiful girl from a "respectable family," fell hard for Jim. "She pursued Jim with a vengeance. If it had been the other way around, I would have had her in the back seat of my Volare in a half hour," Mark says. Jim finally dated the girl after clearing it with Mark, but he was never the pursuer in his relationships. "I knew for a fact he wasn't gay, because I'd come across many gay people. He just didn't care."

It wasn't that Jim lacked passion. "The second love of my life, after my cowboy boots, was a '71 Barracuda with rust holes, a dented primer body, and studded snow tires," he says. Jim paid for the car by cleaning grease flues from restaurant kitchens late at night. Mark and Jim spent every free moment they had on their cars. "When you're a sixteen-year-old boy in the Midwest, your car is your whole world," says Mark. The friends worked for months on Jim's Barracuda. "Jim repainted and restored the back part of an otherwise piece-of-shit car. It was just one small part of that car, and it started to become beautiful and admirable. And Jim parked it in his driveway that night. His dad came home and saw that it had been fixed nicely, and ran right into the back of the car with his brand-new Lincoln."

High school, for Jim, became what he called his "six-hour recovery program." By the time he was into his second year, he drank late into the night and was tired and hungover in the morning. "Jim was a horrific drunk," says Mark, "and every opportunity, he would drink himself into oblivion—or take long walks. That was how he dealt with everything." There was no incentive for him to show up ready for scholastic achievement. Instead, he loved tinkering with tools and cars and spent most of his time in shop class. The one teacher he bonded with was the shop teacher. "He picked me up in his van with shag carpet on the ceiling, smoking a joint on his way to school," he says. By the end of the school year, he had flunked out of half of his classes. The other half were study halls where he'd nap.

The four-year mandatory curriculum stretched into year five. Franklin McCallie, the head principal at Kirkwood, remembers Jim as one who stood head and shoulders above the rest of Kirkwood's degenerates when it came to bad behavior. While other kids would skip class and not do what they were supposed to, "Jim was so *in your face* about not doing what he was supposed to do. We kept saying, 'There are rules you've got to follow. There are things you've got to do. There are grades you've got to make, there are tests . . . ,' and so on. And he just wasn't doing it, which is why I said, 'You're not going to graduate. You can't graduate.'" Jim, cocky and obnoxious as usual, replied, "Frank, baby, I got nowhere to go. I got nothing else to do. I'll be back next year."

The threat of not being able to graduate did nothing to mellow Jim's love for wreaking havoc on campus. He nearly landed in jail shortly after his conversation with McCallie. He stole an industrial smoke bomb and lit it off in the school's north building. It went out as a five-alarm fire, with a half dozen fire trucks and the entire school evacuated. That night Principal McCallie arrived at the front door of the Kennedy house and began to explain to Jim's parents that the school would seek criminal charges to prevent him from coming back to school. "My father grabbed me and dragged me to the door. He looked at the principal, and without skipping a beat, punched me in the side of my head. The principal is looking stunned at my dad, who's explaining how he's really going to take care of me this time." Jim's mother was watching from the hall. "She starts laughing hysterically and says, 'Oh that Jimmy, he's such a kidder.'" McCallie left without another word.

The next day at school, McCallie called Jim to his office. The six-foot-three giant towered over the angry teenager, for whom he felt great compassion. McCallie had a reputation for his kind ways and his great empathy for his wayward students. He would do nearly anything to keep from expelling a student, even one as awful as Jim Kennedy. Jim remembers the conversation as one of the most pivotal of his life. "Do you really want to graduate?" McCallie asked for the second time. "I'm like, 'Yeah.' And he told me, 'Then you need to come to school at 5:30 A.M. and stay until 5:30 P.M., and you're going to clean up all the trash in this school. *You hear me?*'"

Over the next three months, while Mark and his cool friends parked their prized cars in front of the school, smoking and listening to Richard Pryor and Steve Martin comedy routines on their eight-track players, Jim swept up cigarette butts and fast-food wrappers and buffed floors with the janitors. As he cleaned the school that had been a miscreant playground to him for five years, a little flicker of self-confidence began to infiltrate his hardened young spirit. It was a job, and he went diligently every dark morning. He arrived on time, and he did his job well. Instead of throwing him out of school, Franklin McCallie gave Jim Kennedy a purpose. And without that great leap of faith by a beleaguered public school principal, he says, "I probably wouldn't be alive today." When the semester wrapped, "Guess what Frank had for me? A shiny new diploma."

On the last day of school, Elizabeth Kennedy asked her son what he would like for his graduation gift. The two drove to the liquor store, where she fulfilled his request: two cases of Michelob *in the bottle,* because "I was a snob." He cracked one open and headed for campus. In one final act of defiance, Jim drove straight onto the athletic field, spinning in wide circles on the grass and then speeding through all the common areas. "Every teacher at the school knew it was me. No one tried to confront me or stop me." Afterward, he parked the car in his usual spot with Mark and his buddies. "As my friends were leaving, I was handing each of them an open beer. A teacher pulls up, sees what I'm doing, and says, 'Are you fucking insane?' I said, 'Yes, I am.'" The teacher, like most adults who came across him, could see no upside to escalating the situation. He drove away. "Everything I did, every decision I made, was like Russian roulette. My life was such that I didn't care about the consequences."

Jim had little reason to care. At home, he still could not escape the anger and brutality of his father. There was, however, one quality his father had that Jim Kennedy managed to internalize. Despite growing up on a farm in a poor and loveless family, Gerald Kennedy raised pigs and invested in stocks to pay for college. In college he boxed above his weight class. His strategy was to exploit his opponent's confidence. It was his father's single contribution to Jim's "Good" list: When you take a leap of faith, never look for the net below.

It would be the single trait that Jim Kennedy would employ above all others, one that would propel him throughout his life and catapult him into the biggest battle of his life—the fight for what he believed was the future of the United States of America. Even when everyone close to him begged him to give up, when everyone in Washington, D.C., told him his fight was futile, his determination only escalated. "*Never be afraid to fail.* My dad taught me that. That was his one gift to me. That was his one lesson."

4

The Biggest Puzzle

Before the move to California and the stroke that took away his beloved mother, there was a magical time in Jim Kennedy's life. In those early years, the Kennedys lived far outside the St. Louis suburbs, surrounded by enchanting woods with creeks and ponds. Elizabeth Kennedy was healthy, and life was good. It was here that Jim first fell in love with the earth and all of its treasures. To little Jimmy, the hills and rocks and fossils were alive, and they embedded themselves in his explorer's heart. In the desert of Southern California, he obsessed over the neighbor's landscaping rocks: pumice, lava stone, and granite with crystals of sparkling mica that he peeled by the layer. He collected rocks, but like everything good among the Kennedy siblings, his treasures fostered rivalry and jealousy and disappeared along with his motivation to organize and admire his shiny finds.

By the time the family returned to St. Louis, twelve-year-old Jim, with a growing catalog of abuse and discouragement, had lost his curiosity for nearly everything. Still, his deep attraction to the earth remained imprinted on his Midwestern soul. In some of the few happy times Jim remembers with his father, Mr. Kennedy would stop the family station wagon at a road cut, exposing a delectable slice of Missouri geology. There, Jim filled his pockets with rocks and fossils, marveling at the three-dimensional view of history. One of the few topics that captured his imagination in high school was plate tectonics. "I understood the earth was alive—an organism of silica and stone," he says. "I was fascinated." By the time he was a teenager, he was traversing the suburbs of St. Louis, contemplating only how to survive. He all but forgot his love for rocks and hills and streams. His conscious mind did not feel the earth below as she continued to pull on him, nor could the immensely pained teenager imagine that in twenty years' time, the unique geology under his feet would become an obsession that would change the course of his life forever.

More than three billion years before a recycled collection of atoms became Jim Kennedy, during the earth's pre-Cambrian period, the area that is southeastern Missouri began writing its geologic story. Exploding volcanoes blanketed the terrain with ash and lava, their magma chambers cooling over millions of years into mounds and masses of red-speckled granite. Ages later, the sea level rose, and salt water poured into the continental interior. Oceans hosted creatures that built carbonate reefs, and waves wore down river deposits into sandy shores, building a rich landscape later carved and chiseled away by glaciers. All the while, the extreme heat and pressure of the earth's core pushed its way to the surface, fracturing and faulting the rocks above into rugged terrain laced with rich deposits of lead sulfide. This random act of thermodynamics was the very thing that would first bring great riches to the territory that would become known as the Missouri Lead Belt.

It was a Frenchman who first recognized economic ore in what was then French-owned Upper Louisiana at the turn of the eighteenth century. Exploring the wilderness near the Illinois territory while trying to convert the Kaskaskia Indians to Catholicism, Jesuit missionary Father James Gravier encountered a "very rich lead mine," twelve or thirteen leagues from the mouth of the Miramigoua River. It was only a single mention in his journal "on the 10th day of October, 1700," but he noted that the rock was very pure—fully three-quarters metal. Stories of the find, enhanced with rumors of silver and gold, traveled all the way to Paris. Hearing of possible riches, Francois Renault, a Parisian banker and favorite of French ruler Philippe I, got a job as the director-general for mines with the Company of the West. The company was granted a monopoly by the French government to control trade between France and its Louisiana and Canadian colonies. In Canada, the French traded in beaver skins. In the Louisiana Colony, they would trade in gold, silver, and other precious metals.

In 1723 Renault set sail from Paris with a crew of two hundred men. In addition to provisions for the long journey, the ship's cargo was heavy with mining tools and bricks to construct smelting furnaces. In Santo Domingo (modern-day Dominican Republic, then a French colony and way station for vessels sailing to Louisiana), Renault purchased five hundred African slaves, for a total crew of seven hundred. From there the ship traveled along the swampy coastline of Louisiana and then northward up the great Mississippi River. As his mandate was to discover precious metals, Renault sent men out to look for ore along the way. His knowledge of geology was clearly lacking, and his men returned with discouraging tales of endless earth. The floodplains of the Mississippi were rich not in precious metals but in fertile soil deposited over thousands of annual flood cycles. By the time Renault reached the area of modern-day southeastern Missouri, he could barely manage his crew aboard the ship's tight quarters. They disembarked into a French settlement on the Mississippi, bringing the first African slaves into the territory. The settlement, named Kaskaskia after a regional tribe, became known as the "metropolis of the Mississippi."

From Kaskaskia, Renault continued his exasperating search for ore. Deep into the surrounding wilderness went the parties. Weeks turned to months as nothing in the wilds gave any indication of the wealthy chunks of rock they were so desperately seeking. When things seemed completely hopeless, one party returned with samples of a shiny gray mineral, heavy to hold and formed of bulky, cube-shaped crystals, just as Father Gravier described. The lead sulfide ore (dubbed galena by Pliny the Elder in 77 A.D.) blanketed the forest floor with a lustrous sheen. While silver or gold would have been a far more lucrative bounty, the lead was still valuable, especially in such quantities. Highly malleable and noncorrosive, it was an excellent metal for all kinds of uses, from glass and ceramics to bullets and buckshot.

Renault's men used shovels and axes to dig shallow pits. As the holes became deeper, one man would lower himself into the pit, and one man would stay above, running a winch with a bucket attached to pull up the ore. Worthless rock was separated from galena by "jigging." The method used a long wooden pole with a fabric sieve full of crushed ore tied to one end. The bag of ore was placed into a stream, while the opposite end of the pole rested on a log or wooden horse. Then a miner moved the sieve up and down in the creek, causing the heavy galena to sink to the bottom of the sack. The ore was then "smelted" (a process that separated the metal from its ore mineral) in brick furnaces. It wasn't particularly difficult. At just 325.5 degrees Celsius, the chemical bond between lead and sulfur breaks. Sulfur atoms grab oxygen molecules in the air, sending up volatile sulfur dioxide and leaving a puddle of molten lead below.

Production soon reached 1,500 pounds of lead per day. Mules and wagons carried the metal to the town of St. Genevieve on the Mississippi. When carried by packhorses, the lead was not in "pigs"[1] but in horseshoe-shaped collars that the horses could wear around their necks. From there it traveled south by ship, and finally to France. Unfortunately for Renault, mining in the Upper Louisiana territory turned out to be much more difficult than he bargained for, and his endeavors were far from lucrative. Brutal winters—the likes that the French and especially the Africans had never seen—cut the season short. Canadian fronts froze the ground, making digging impossible. In the summer, torrential rains filled the shallow pits and shut down all operations. Renault also complained that he had no protection from the warring Osage Indians, who still controlled much of the surrounding wilderness.

An unexpected turn of good fortune came about for Renault in St. Philippe, a nearby village he built to produce food for his men. Because the soil was so fertile, agriculture became Renault's most successful enterprise. St. Philippe farms produced boom crops year after year, and Renault sold the surplus to French settlements all the way down the river to New Orleans.

Renault continued mining operations until 1744, then transferred his holdings to the French government and returned to France. In 1763 France lost the French and Indian War to Britain. Afterward only small, independent mines—something more along the lines of subsistence farming—were in operation. Having already relinquished Canada, and in order to keep Upper Louisiana out of the hands of the British, King Louis XV secretly ceded the west bank of the Mississippi River and surrounding French territories to Spain in the Treaty of Fontainebleau.

It took almost seventy years after Renault's departure for the deep riches of the Lead Belt to see a second wave of exploitation—this time in the character of a bankrupt Virginia mine owner named Moses Austin. In 1796, after learning of the lead-rich land out west, thirty-three-year-old

Austin left his home in Richmond on horseback with his servant and a pack mule, hoping for better luck out west.[2] The thousand-mile journey, he would later recount to Henry Schoolcraft, a geographer and writer, was an "extraordinary effort of hardihood." The land was wild the entire way; rarely did they encounter any structure built by human hands. Toward the end of their harrowing journey, they were lost in a horrendous snowstorm, ran out of food, and ended up sixty miles off course.

Finally, in St. Louis, Austin crossed the Mississippi by ferry and made his way southeast to the legendary lead deposits. Knowing the value of making an impression to local Spanish officials, he donned an embroidered blue cape lined in scarlet. He rode on his best horse, followed by his servant, past a few ramshackle houses scattered along the river and up a dirt road to the lieutenant governor's home. After Austin showed the governor his official papers, announcing the Spanish connections he'd made before leaving Virginia, the governor immediately offered his residence to the stately American, stating that it was the only house in town that could offer him suitable accommodations. After several weeks Austin, enthralled with what he discovered, convinced Spain to grant him over 4,000 acres to develop a mining industry in the territory. In return, he swore allegiance to Spain, and promised to work on settling the area for the Spanish crown.

Austin invigorated the mining industry straightaway.[3] Too much time was being wasted by men digging new pits each time one became too deep, so Austin had his men dig seventy to eighty feet down, then follow the linear galena deposits by digging horizontally, extracting the ore along the way. "Drift mining," as this technique was called, halved the time spent digging. For the first time, the miners could work underground, which allowed them to continue mining in all kinds of weather. Austin then erected the area's first reverberatory furnace, an ancient but much more efficient smelting method that tripled yield. By 1802 Austin's smelter was the only game in town, and he processed galena for the entire mining area. He named his settlement Petosi, after Petosí, Bolivia, a city famous for its silver mines. Then he developed a shipping center he called Herculaneum, after the ancient Roman city, apparently because the limestone bluffs along the Mississippi reminded him of an amphitheater. Giant brick shot towers built on the bluffs produced lead bullets and cannonballs. Over the years, Austin expanded his businesses, continually employing a crew and staff of forty to fifty—some paid workers and some slaves. For four straight years, starting in 1805, Austin's mine produced over 800,000 pounds of lead. He built sawmills, a general store, roads, bridges, and ports, and a lavish house he named Durham Hill after his hometown of Durham, Connecticut.

Austin's fortune was not all good. In 1802 about thirty Osage Indians attacked his home. The French settlers—who'd always despised Austin—refused to intervene. He was able to keep the Indians from destroying the mine workings, his mill, and his home, but one man was killed and a woman was kidnapped. Afterward, Austin repaired an old cannon to ward off future invasions, but the threat of another Indian attack and the sting of his French neighbors' abandonment were constantly on his mind.

In 1803 the Lead Belt came under the jurisdiction of the United States as part of the Louisiana Purchase. Austin continued to build his lead empire, uninterrupted by the new flag. In 1816 he wrote about the galena deposits in Missouri, "From the state of facts, it is impossible to estimate their value; that they are immense, no one will deny; nor can they be exhausted for ages, if properly managed." With his great wealth, Austin founded the Bank of St. Louis, but shortly after, his life was brought crashing down by the New World's first economic collapse, the Panic of 1819. Rising to the position of a great Father of Industry had come at a price. Austin had many enemies, and he was embroiled in an exhausting lawsuit. On March 11, 1820, he was arrested and put in jail for failing to pay his debts. The government seized his most prosperous mine. Upon his release, Austin and his son, Stephen F. Austin, left for Spanish Texas, hoping to remake their fortune. With a land grant from Spain, Austin planned the first American settlement there. He never made it. He died of pneumonia in 1821, $15,000 in debt, leaving twenty-four-year-old Stephen to colonize Texas.

Another four decades would pass before a third wave of Lead Belt exploitation brought together a curious coupling of characters. Together—with great ingenuity and perpetual doggedness—a lawyer from Utica, New York, and a dentist from Michigan would bring great riches to the newly founded state of Missouri.[4]

Although Moses Austin had left the territory more than half a century before, the lead belt's riches remained legendary across the country. In March of 1864 a group of New York investors who knew nothing of the mining business formed the St. Joseph Lead Company. For all intents and purposes, it was just another of the many stock market schemes popping up during that era. The company's incorporators gathered money from investors, purchased almost 1,000 acres of lead territory, and hoped to turn a quick profit. It was not a sensible plan. They had no guarantee of the lead content in their purchase, and did not know how to extract it, if it were there at all. The company was instantly $75,000 in debt for the land, with most of its stock sold and little cash on hand.

The complete muck-up that was the St. Joseph Lead Company must have been quite shocking to a middle-aged lawyer by the name of J. Wyman Jones, who accepted a block of stock as payment from a client. After seeing a notice in a New York newspaper, a curious Jones decided to attend the company stockholders' meeting. It was June 13, 1865. The company was barely a year old. A gloomy picture emerged, and so many questions from investors went unanswered that the meeting was adjourned. At a follow-up meeting, perhaps because he seemed to be the only person actually interested in the mining operations, the board invited Jones to assume the company's presidency.

Growing up in the forests of New England, Jones had always loved the outdoors. He quickly accepted the challenge and left his Utica law practice for the Missouri wilderness. While his physician gave him advice to find a job that would offer fresh air, the stress of his new position was probably not the prescription the doctor ordered. Jones's first report to the trustees told of serious problems. An immense drought the previous summer left streams dry, making jigging impossible. The following fall, the company's operations were raided by the Confederate army, led by General Sterling Price (Missouri was a contested border state in the Civil War). For two months, the miners refused to return to work. The war was over, but winter came fast and hard, followed by torrents of spring runoff that flooded the mines.

Jones kept at it, but the following year, in 1866, it was apparent that lead mining was a losing proposition. He wrote to the shareholders, "It will be readily seen that the labor involved in blasting, moving, breaking, crushing, jigging and smelting ores was quite out of proportion to the value of say 2,300 pounds of lead per day at five cents a pound, and that we derived little encouragement in the way of profits." In all, the mines could only stay in operation for half the year, far too short a time to cover expenses. Getting the metal out of the territory was another challenge. Roads through the woods became deep, muddy trenches in which the heavy wagons sank up to the axles. Hauling slowed so much that it was impossible to secure enough mule teams to handle the job.

A slight turn for the better came later that year, when Jones hired a number of Cornish miners, so-called Cousin Jacks. The workers left Cornwall, England, after their famed tin and copper mines went broke during the mid-1800s when global competition dropped the price of copper. With their highly skilled labor, they also brought mining nomenclature that was quickly adopted in the Missouri territory. Pits were called "shafts." Horizontal tunnels were "levels." Depending on the angle, "winzes" or "raises" joined two tunnels, and drainage tunnels were "adits." When the men were deep down a shaft, the Cornish code of signals was used to communicate between the hoisters and the men below. The Cornish pump cleared the mines of rainwater and groundwater. Canvas helmets with oil lamps illuminated the rock faces deep underground. And to make the miners' days more comfortable, they all carried a special tin bucket with their midday meal. This well-thought-out contraption came with separate compartments; tea was heated in the bottom, a meaty Cornish pastry sat in a middle tray, a second tray held a miner's dessert, and a metal teacup was attached to the bucket's lid.

Even with the additional know-how of the Cornish miners, St. Joseph Lead had yet to make one penny of profit. Jones may have felt like quitting, but instead, after the company's annual shareholder meeting in October 1866, he decided to take a break with his family back in Northampton, Massachusetts. In his comfortable summerhouse, luck finally came to Jones and his beleaguered company in the form of one Dr. Charles Bunyan Parsons. Parsons, a former captain in the Union Army, had been forced to resign because of poor health. Like Jones, Parsons was persuaded by his doctor to find a career where he could be outdoors—which was how he ended up, with no mining experience, at the helm of a Northampton mining company. By the time he encountered Jones, the mine was closed, and Parsons was out of a job.

Jones and Dr. Parsons bonded over their passion for and naïveté in mining. Jones told Parsons about his many troubles, but said he was still enthusiastic about the tremendous possibilities in the Lead Belt. While Parsons had little more knowledge than Jones, Jones asked Parsons to investigate the Missouri mines and report back to the company's board of directors in New York.

The thin, narrow-faced dentist who arrived in the wilderness of Missouri was a natural. He looked over the mines, gathered as much information as he could about their workings, and took copious notes. Back in New York, he presented his findings to the board. Because he grasped the problems so effectively, he was quickly offered the job of resident manager.

In 1867 Dr. and Mrs. Parsons began the long trip to Missouri. A month later, they arrived at the pitiful mining operations, which consisted of little more than an outpost with thirty log cabins. The settlement was called Bonne Terre, after La Terre Bonne—"The Good Soil"—the name that the French gave the area in 1720. The "works" at Bonne Terre consisted of one small crushing mill, one set of Cornish rolls where rock was crushed and processed by a few hand-operated jigs, two iron drums, and three reverberatory furnaces. To increase production, Parsons instructed his men to dig the tunnels at an angle so they were not so apt to flood. Parsons purchased more pumps to keep the pits dry enough to work. More furnaces were installed, and Parsons designed a mechanical jig that took over the one-man job. Parsons earned a reputation for being enthusiastic and wise, caring and kind, inspiring to his men and dedicated to his company.

At first the St. Joseph board was less than thrilled by Parsons and his lofty expenditures. But before long the mine works processed enough lead to make a profit and pay shareholders a small dividend. Elated, the company sold $100,000 worth of bonds. By January of 1868, less than a year after arriving, Parsons had brought the company so much success that, with the exception of money owed the stockholders, St. Joseph Lead was debt-free.

Parsons was not satisfied with these modest gains, however. Reading a magazine article about a new invention called the Lechot diamond tunneling drill, he became obsessed. The drill had been invented to help excavate a tunnel through Mount Cenis on the border of France and Italy. Dynamite placed in the drill holes blew open a tunnel between the two countries. The article described how the steam-powered drill—with a hollow circular drill bit covered in Brazilian diamonds—descended deep into the rock, spinning at 300 revolutions per minute. What came out at the surface was a "drill core"—a long cylindrical sample of the geology below.

Parsons wasted no time convincing Jones that the diamond drill could be of great use in their mining operations. Rather than digging large pits to look for ore beneath the surface, the drill could prove straight away whether digging would be profitable at all. Renting the drill was pricey, so Jones pitched the idea to the board, hoping that they would approve the necessary funds to bring the drill to Missouri. The board refused. There was ore literally everywhere, they argued; it would take years just to process what had already been uncovered. Fortunately, the company's treasurer had more foresight. He agreed with Jones that although much ore had already been uncovered, there was no reason not to look for more.

On March 5, 1869, the rented drill arrived at the mine. It was a beautiful sight. The former dentist marveled at the machine as it hollowed out its first core of the Missouri geology, exposing layers of geologic history in smooth rock cylinders. For months, Parsons reworked and repurposed the massive machine until he had the perfect tool to reveal the wealth under the surface. The drill was an economic windfall for St. Joseph, starting a revolution in the mining world. In August 1870 the editor of the *Washington County Journal* heralded Parsons's brilliance:

> The mining fogies of our county would find their mouths shut (if 'twere possible) were they to visit the St. Joe mines. Major Parsons, a gentleman educated in what schools and practical experiences could impart, is placed in a position in which the Lechot Diamond Tunneling Drill fails to perform certain requirements. The profit side of the annual statement will be injuriously affected unless a certain set of new operations can be performed by this drill. Hero Major Parsons steps in, and by taking the

> Lechot drill apart, supplies by his own genius the combination of movements that adapt the Diamond Drill to work in all the required positions, giving an important agent to facilitate shaft sinking in the future. Oh! Ye pick and shovel men, of *small* and feeble ideas! How long would the best of you be in preparing yourselves to reconstruct a Diamond Drill, or to prepare your brains to manage such a property as the St. Joseph Lead Mines?

St. Joe expanded throughout the territory, absorbing all of the smaller mining companies in the Lead Belt. Two hundred and eighty underground miles of haulage track carried ore under the power of thousands of mules, many that were born and died without ever seeing the light of day. The bond between the community and St. Joe was so strong that community members often referred to the company as Uncle Joe. Houses were built for the employees, with porches and shade trees. A company store charged workers only 10 percent over wholesale. St. Joe raised cattle for beef and dairy, selling it at only a modest profit to their workers. A U.S. post office popped up in Bonne Terre. There were churches and schools, all built with the help of the company. Jones, still at the helm of St. Joe, was known as a fair and caring president, dedicated to the needs of his employees and their families.

Jones died on October 27, 1904. He was seventy-two. During the nearly forty years he spent at the helm, St. Joe grew from a wilderness outpost to a company that distributed $1,774,000 in dividends to its shareholders. The operation boasted nearly $2 million worth of permanent infrastructure, and a thriving community of 5,000 in Bonne Terre. Six years later, Dr. Parsons, the good-natured engineer who changed the world of mining with his diamond drill and many other inventions, died at his home on the Mississippi River.

St. Joe, like many other companies in the 1930s, was gasping for air as the Great Depression strangled American enterprises. Rather than close operations, in 1932, the company shut the mines for all but one week per month. Dwight Jones, son of J. Wyman Jones, was now the president of the company. Like his father before him, he made taking care of his employees of the utmost importance. During the summer the company provided areas for employees to raise food for their families. During the winter, it let its employees cut wood for heating and cooking from company land.

Fortune came again to the Lead Belt as World War II began. Prices for industrial commodities skyrocketed. St. Joe had more than 3,000 employees by the end of 1945, and the area boomed. St. Joe built a clubhouse with seventeen guest rooms, where young engineers could come and stay "until such a time as they were established in the community," remembered a former miner in a 2014 interview. "I ate there in the '40s especially, quite often would eat lunch. Fifty cents, you could get a full meal including dessert and drink and everything. So, in every way they attempted to make the life easier for their employees."

But by the end of World War II, the great Lead Belt, which spanned seven counties and once contained the largest and most highly concentrated reserves in the world, had been plundered to near extinction. An estimated 8.5 million tons of elemental lead had come from what, in retirement, would come to be known as the Old Lead Belt, in mining operations that produced around 250 million tons of mining waste. The waste rock, commonly known as tailings in other mining operations, was called "chat" in the Lead Belt. Chat piles, hundreds of feet in elevation, were playgrounds for Lead Belt kids and a beloved feature of southern Missouri topography.

Knowing that in an area of such riches, there were sure to be more, St. Joe formed a geology department to search for new deposits. The team headed west of the St. Francois Mountains, since all of the deposits to the east of the range were depleted. There were no known surface deposits in the area, so the geologists used modern instruments to help them guess what was underground. Over the next decade, the St. Joe team discovered massive amounts of galena in a deposit they called the Viburnum Trend. They found zinc sulfide ore, which was also well known in the territory. And then they encountered something quite unexpected. Using an instrument called a magnetometer to uncover rocks that were magnetic, they found a signature so strong it could only mean one thing: iron.

On a crest called Pea Ridge, in the town of Sullivan, Missouri, St. Joe began drilling operations to find the ore body that was sending their magnetometer off the charts. For more than 1,000 feet, the cylindrical core surfacing from the hollow diamond drill pipe showed nothing but worthless sedimentary rock and ancient volcanic deposits. Then, from 1,300 feet deep, what arrived on the surface, to the delight of the geologists, was a core of reddish gray hematite. They drilled another 600 feet and found deep black magnetite—the purest form of iron ore. There was no way to know how much deeper the ore body went. Many more samples collected over several months teased out a massive crescent-shaped ore body, half a mile long—the largest magnetite ore body ever found in America. The geologists, still having no idea of the ore body's depth, guessed conservatively that the Pea Ridge iron deposit held 200 million tons of pure magnetite.

With little infrastructure to deal with iron ore, and no clue how to process it and market it, in 1957 St. Joe partnered with the world's largest steel producer, Bethlehem Steel. The two giants formed a new company called Meramec Mining Company, after a nearby river, and went to work. The geologists, engineers, and miners who broke ground in the rich St. Francois soil were a lively bunch. One worker took it upon himself to write a weekly newspaper called the *Mangled Iron-er*, highlighting the mine's drilling progress with a touch of tongue-in-cheek safety advice: "Don't jump to conclusions about safety, you may slip,"[5] and "BE EXTRA CAREFUL—You and your equipment may both be rusty."[6] The *Mangled Iron-er* was also the source of Pea Ridge gossip and goings-on: "Jack Emery has leased the floor and foundation of the old utility building and turned it into an Ore House, and he is right out front in the open with it."[7]

"At the time, working at Pea Ridge was one of the best jobs around," retired miner Larry Hoffman told the *eMissourian* in an interview in 2013.[8] "They were hiring like crazy in the 1960s. All the guys were best of buddies. It just seemed like everyone had a heck of a good time." The men went by nicknames only and played practical jokes all day, Hoffman said, like nailing lunch boxes to tables. "I've never met a bunch of people who were so nice . . . They're good guys and good workers. They were brought up to work." Following the lead of St. Joe, Meramec paid well, and the benefits were great. Employees got thirteen weeks of vacation per year. "It was a wonderful place to work," says mining engineer Tom Gallagher, who followed his father to Pea Ridge straight out of high school in 1968, working summers until he started full-time after college. "Heck, everybody in this little town worked there or someone in their family worked there. It was the only major employment around here unless you drove into St. Louis County." Gallagher would grow to feel like the mine itself was a part of his family, but the first time he went down the shaft in the "cage," he was sure he'd made the worst mistake of his life.

There was a hazing of sorts for new guys on their first shift underground. At a union meeting the night before at a local watering hole, "the boys were pumping me full of all kinds of crap about working underground," Gallagher remembers. "When I caught the cage that first day, I didn't like it at all." The hoist man turned up the speed as fast as he safely could. Hungover, Gallagher tried not to throw up as he watched the shaft steel and timbers whiz by. Then the hoist operator hit the brakes—a little harder than normal, because he knew there was a new guy on board. "I had all this survey equipment, and my knees almost buckled. I staggered to the edge of the cage and caught myself." Then he had to climb up a forty-foot ladder tied together with number 9 wire and dangle a plumb line for the surveyor to shoot. "I'm hanging on for dear life, and the sweat is just pouring off me. Everyone just sat there grinning. I thought, Oh, Gallagher, you've make a HUGE mistake, you really don't want to be doing this." It wasn't love at first sight. "But after I spent some time underground, I got to where I *loved* it. That's where I wanted to be."

The tunnels, lit only by the miners' headlamps and machines, were a comfortable twenty feet high and wide. Some of the "rooms" had thirty-foot ceilings. The men blasted into the black magnetite with dynamite, breaking it loose into millions of pieces of rubble. "In the early days, everyone was young and bulletproof and thought they could do anything," Gallagher says. "I was a young man, and it was kind of a big boy's dream. All of the equipment was big and loud and noisy, and there were explosions going off. You could play in the mud and momma didn't paddle your tail when you got home."

By all accounts, working at two thousand feet below the surface was a dangerous endeavor. Gallagher becomes somber when remembering the sixteen workers who died at the mine. Once, a close friend blew himself up when he forgot to cut a lit igniter cord attached to his underground vehicle. The Missouri Health Administration inspected the mines, but in those days it didn't have the ability to issue citations or implement safety standards.

In December of 1977 a global oversupply of steel gutted iron ore prices, and Pea Ridge closed. A thousand workers lost their jobs. It was the beginning of the end for Bethlehem Steel. The great symbol of American steel manufacturing left the partnership with St. Joe and began a decline from which it would never recover. In 2001 Bethlehem Steel went bankrupt—its final holdings absorbed by the Indian-owned, UK-based corporation Mittal Steel.

Tom Gallagher and mining geologist Larry Tucker were part of a skeleton crew of fifty who were hired back. In 1979 St. Joe resumed operations at Pea Ridge on its own, with mostly new, nonunion workers. Gallagher was incredibly relieved; leaving his mine was unimaginable. "When the mine started back up as Pea Ridge Iron Ore Company in the late seventies, I took it upon myself to landscape the place. I planted trees around the office buildings and just tried to make it a nice place to come to work." For a few years Pea Ridge did very well, remembers Tucker. "We had four or five years, and some of them were pretty spectacular. Then the market went down and customers went away. There was no capital. For twenty years, we got no new equipment underground. That was hard. It's a hard business."

In 1981 St. Joe, the benevolent company that built so many southern Missouri families and the greatest lead producer in the world, could no longer compete in a global economy. Profits dropped more than 40 percent. Fighting off a hostile takeover by Canadian alcohol distillery the Seagrams Company, which was looking to branch out into other industries, St. Joe's board agreed to let the company be absorbed by the enormous California-based Fluor Corporation.[9]

When Missouri State geologist Cheryl Seeger first set eyes on the tremendous body of magnetite illuminated by her helmet lamp, Pea Ridge was limping along financially under Fluor. It was the late 1980s, and at the time the U.S. government was looking for new deposits of strategic and critical minerals. Because of a massive find of copper and gold recently discovered in an Australian iron mine, Seeger and her colleagues were given the task to nose around and see what they could find associated with the Pea Ridge magnetite.

On her first day at the mine, Seeger stepped into the cage and descended down the 2,000-foot shaft laden with the tools of her trade: a clipboard with Mylar graph paper to sketch the rock face, colored pencils to mark the different rock types, a Brunton compass (which became completely ineffective next to the massive iron ore deposit), steel-toed boots, and a helmet like the miners wore. She felt no fear, only exhilaration. Underground, massive fans blew fresh air into the cavernous shafts and tunnels, which the warmth of the earth kept at a balmy 70 degrees, year round. It was glorious. It was also the most puzzling geology she'd ever encountered. "I remember trying to start mapping and just spending most of the day going, 'What is going on here?'"

The management assigned Larry Tucker to work with Seeger. Pea Ridge hadn't had a full-time geologist on staff for a while, and Tucker was happy to have Seeger's insight. "We spent a lot of time hanging around the edges of the ore body," Seeger says. "We'd sort of sketch the line of the face and color in what the rock type was, and then draw in any fractures."

It took Seeger a month to catch her stride. "Then we started to feel our way around better," she says. The most fascinating part was that surrounding the massive magnetite were enormous pipelike structures called breccia pipes, fifty feet in diameter. Inside these, the rock looked like Italian mosaic pottery. The pipes were four hundred feet from top to bottom that they could see, and certainly went much deeper. Seeger knew that breccias were the result of enormous explosions that happen when superheated fluids explode underground, fracturing the rock they burst into. "So we got to work out the whole puzzle, which to me is the best part of geology. And you have the biggest puzzle to work on, which is the planet earth. Where else do you get to three-dimensionally walk through geology than an underground mine? It's totally unique."

"Cheryl is a geologist's geologist," says Tom Gallagher. "She took such an interest in things and took copious notes on everything. She'd carry her little geologic hammer around, pecking on the pillars [areas of hard rock left in place to support the weight overhead] and the walls and everything.

Everybody liked her. She was a real professional, but she didn't mind putting on her boots and going out and getting dirty."

Seeger banged away at the strange formations and collected samples to send to the U.S. Geological Survey (USGS) for analysis. The results were quite impressive. The breccias contained a mineral called monazite, and the monazite was chock-full of rare earth elements with an average of 12 percent rare earth oxides—far greater than the 1 or 2 percent commonly found in known rare earth deposits. In the late 1980s, there weren't a whole lot of applications for rare earth elements. They were just beginning to be used in medical applications, as additives to petroleum, and to enhance the capability of lasers. So even though there was very little demand, "We did look at it in terms of being a potential resource," says Seeger.

Unfortunately, with metal prices in the toilet in the 1990s, the rare earths meant nothing to the Fluor Corporation, which, by then, wanted out of southern Missouri mining altogether. Fluor spun off its lead and zinc holdings to various mining companies. The Pea Ridge iron deposit went to Big River Minerals of Missouri for $13 million.[10] America's largest and purest underground iron mine limped along for eleven more years, finally collapsing to the pains of outsourcing, globalization, and costly environmental regulations. In 2001 the mine went dormant, its caverns filled with groundwater, its lift rusting in the humid summer air. Below the surface, a colossal body of ore sat waiting to be resuscitated by yet another curious player in the story of Missouri mining: a man by the name of Jim Kennedy.

5

The Recruit

Having a Kirkwood High School diploma from the class of '82 did nothing to give Jim Kennedy any confidence in his future. Living at home with his father had turned him into an empty, angry shell. He cared about nothing. "The truth is, I wanted to die. My problems and my environment were so bad, I was actually passively seeking my own demise. I knew I was incompatible with this world. I had no future. I never ever did anything." Jim tended to get along with people—at least those that weren't his family—one-on-one. "But as soon as there was a crowd, I didn't like being there." School ingrained into him that he was little more than a future government statistic. "I took general math in my senior year. It was the minimum requirement for graduation. And I got a D-minus; that's adding, subtracting, multiplying. Reading was slow and difficult, so I avoided it. Writing was even more difficult. Spelling was a mystery, and I never, ever figured out how to properly construct a sentence. I had no place in the real world."

There might be nothing for him in ordinary society, but Jim's imagination took him to a world of darkness and shadows, where there could be a place for him after all. He would join the army. He would use his military training to become a mercenary—a hired assassin. "I was going to kill people for a living. Then one day my employer would probably whack me, and the lights would go out," he says. He imagined this scenario, and it gave him some peace. It was the early 1980s. Television news blasted scenes from Communist uprisings in Nicaragua, Guatemala, and El Salvador. "In my mind, I'm working for some shithead, whacking people in Central America. And I was perfectly happy with that."

Jim told none of his family members he was leaving to join the army. "I hated my dad so much, I wouldn't give him the dignity of telling him I'm leaving." Only three of his friends knew. Mark Haselhorst remembers driving him to the recruiting office. "It's all okay. I'm getting out of here," Jim kept saying. Mark knew that things were so bad for his friend that he couldn't stay in the Kennedy house anymore. And after seeing how horrible his life had been for so long, Mark wasn't surprised by his decision: "I think he did want to go kill people in the army."

When they arrived at the recruiting office, says Jim, "The recruiter sees me, and he's all happy." It was during the massive military buildup of the Reagan administration, and recruits were in short supply. "He looked at me like, 'Woohoo, fresh meat.'" Jim passed a physical with flying colors. Next, they gave him an aptitude test. Jim immediately knew he had no chance of passing. "I had D-minus skills, and that was generous. But when I took the aptitude test, the recruiter came back and told me, 'Wow, you scored very, *very* well. You could pretty much have any job you want.' I stepped closer to him, put my fingers into the shape of a gun, and placed it at his temple. I said, 'I want the job of looking the enemy in the eye and pulling the trigger.'" It took a full twenty seconds for the recruiter to recover and to admonish the psychotic-looking kid staring him dead in the eye: "Never say that again in this man's army, or you can't be in this man's army."

Jim wasn't joking. He wasn't even trying to seem tough. "I was really looking for a way to be eliminated. So I said, 'What's the most dangerous job there is, something that would put me behind enemy lines?' And he said, 'Special Forces.'" Jim asked if he qualified, and was told, yes, he did. Jim went back to the parking lot and told Mark that he'd just signed his life away. "I remember seeing how scared he was, but he was laughing about it."

Jim had a few months to kill, so he stuck to his normal routine: drinking heavily, carousing with friends, and getting into trouble. He was still close to his mother and commonly drove her to her weekly appointment at the J. C. Penney hair salon. "I'm bored out of my mind waiting for her, and I'm with my friend Roger Hershey. We're making fun of how bad security is in the store."

"We could steal the biggest thing in here, and they won't even pay attention," he told Roger.

The two heaved a giant Persian rug over their shoulders and walked casually out the door. They were vindicated. Store security sucked. Nothing happened. "I thought the game was over, and that we would just leave the rug on the sidewalk outside the store," Jim says. While Jim went back to the salon to collect his mother, Roger crammed the carpet into Jim's car. When Jim came back to the car, his disabled mother on his arm, "Roger's getting arrested and telling mall security that it was my idea." After being booked at the local jail, the two were released and given a date to show up for a trial.

A few weeks later, Jim stood in front of a circuit court judge, a belligerent eighteen-year-old with a plan.

"Where's your lawyer?" the judge asked him.

"Don't have one."

"You better get a lawyer, son. This is very serious."

"You can do whatever you want to me, but you see this? These are my induction papers to the military. If you want to pursue this charge, I'm not going to the military."

On the bench was Jim's police file, with "all the beautiful shit I've done." The judge seemed to muse it over for a while, grumbling. He didn't particularly like any of his options for dealing with the punk teenager.

"You're going in?" the judge asked after several minutes.

"Yes."

"Okay. You go in, this whole thing goes away."

There was no direct order from the judge to make Jim follow through. "But it was just fine with me, since I had no other options," he says. So he went home and grabbed a handful of things to take with him. He tried to sneak out the back door through the laundry room, hoping not to encounter anyone. Instead he found his mother, who was attempting to do the laundry. She looked at him. He knew her vision was so bad that she would not be able to see the guilty look on his face.

"Mom," he said, "I'm leaving."

"Are you ever coming back?" she asked him.

"It floored me. I couldn't believe it. *She knew*. I just told her, 'I don't know.'"

If there was the tiniest soft spot in his hardened heart, it was for his mother. And a tiny part of that tiny spot hurt a little, seeing Elizabeth Kennedy's sad, empty eyes. Without another word, he was out the door, with no intention of ever returning.

The military was a shock. Standing only five foot six and weighing only 130 pounds, Jim Kennedy had never been intimidated by anybody or anything—until then. Greeting him at Fort Benning, Georgia, in the summer of 1983, was the six-foot-eight-inch Sergeant Newsome—"Gruesome Newsome"—with perfect black skin, massive muscles, and the most enormous hands Jim had ever seen. "If he had taken another career path, he would be a male supermodel." Newsome looked down at Jim with seething contempt for the bony recruit. And for the first time in Jim's life, he felt fear. "I'd never looked into someone else's eyes and been intimidated. It was just like all the movies. You're fresh, and they hit you with the fear as hard as they can." It was the first time he had second thoughts about joining up. "I thought, What the fuck did I do? My universe is not in balance."

Strangely, it wasn't long before Jim took a tremendous liking to Newsome. "He was great. He was an awesome person," he says. Another drill sergeant immediately earned Jim's respect as well. "The guy was such a badass motherfucker. He was so strong that he was the top competitor in the army's weightlifting team." The drill sergeant trained people at Harmony Church in the old World War II barracks—Fort Benning's training facility with the reputation for being the worst place in the army. "The intimidation was incredible. I never experienced it before. There was obviously an adjustment period."

In short order, Jim was in love. The military was paradise. It was about pushing himself to extremes, and he thrived. The yelling, the abuse, the endless physical challenges were wonderful. "The only thing I didn't like about the military was running." Jim's short, muscular legs were heavy under the pull of gravity. Unfortunately, Newsome and the other drill sergeant loved to run, and they held a camp record for the fastest two- and five-mile runs, which they would do with raw recruits. Jim suffered horribly. "My body wasn't built for it. I had shin splints and tendonitis in both legs." The shin splints became so bad that Jim could feel calcium forming over the fractures in his legs, but the pain did nothing to slow him down.

He finished basic training as platoon leader. From there, Jim went straight to advanced individual training to learn how to shoot and all of the other basics of soldiering. His next stop was airborne training, which was a breeze. Jim and three other Special Forces inductees enjoyed proving they were tougher than everyone else, especially the recon marines who trained with them. "The way you proved stuff was by being self-destructive and outperforming everyone. We're in Airborne School, and we're training with recon marines, and we're all in the same barracks. The highest-ranking guy was a marine telling us what to do. We're like, 'Fuck you. We're going out drinking.'" In the morning, Jim and his three cohorts would outperform the marines, just to humiliate them. "You had to do sixty sit-ups, and so we did two hundred in the same time period. We maxed everything while pumping alcohol out our pores. We're literally steaming off liquor through the day and then jumping right back and doing it again the next night."

It wasn't long before the monster of defiance living in Jim Kennedy's heart emerged to combat the authority of his superiors. Rules did not make sense to him and his fellow degenerates. And because they didn't like the rules on the base, they decided to rent a dirt-cheap apartment off base, which was completely against the rules. They drove onto the base every morning, half inebriated on cheap beer they bought on base, made a bunch of noise, and woke up the marines. "So this guy got really pissed, and decided he's going to narc us out," says Jim. The next morning, before the sun came up, a companywide reveille was called. Every recruit was standing at attention when the commander came out.

"It has come to my attention that three or four trainees have taken it upon themselves to move off base," he said. "I would like those trainees to identify themselves right now."

Jim got up front and center and executed his maneuvers to perfection. "Which, in a way, is actually flaunting, like saying, 'Fuck you,'" he says.

"Where are the other three?" the commander demanded.

The three others ran up and stood at attention next to Jim. "They're scared shitless, and I'm there beaming with confidence, staring this fucker down like, 'What are you going to do to me?' But the problem was, he hadn't figured out what he was going to do to me, and he could see I was looking forward to whatever it was." The life skills Jim learned as the son of a raging and abusive father gave him the ability to outplay his commanding officer. "I knew most people enjoy transmitting fear with very little action. Most people are afraid to do anything." In the end, there was no punishment. Jim and his buddies weren't even forced to move back on base. "It was totally bizarre," he says. "We were untouchable."

After six months of nonstop basic training, Jim was given leave. He had never planned to go back home, but with nowhere else to go, he took his older brother's offer to drive him back to St. Louis. Things at the Kennedy house were as dysfunctional as ever. After his clandestine departure, Mr. Kennedy hadn't even noticed he was gone for six weeks. When he finally did, he asked Jim's oldest brother Jerry Jr. where Jim was. Jerry Jr. collapsed in hysterics. Mr. Kennedy was furious. Somehow, he composed himself enough to act cordially to his third son when he unexpectedly returned. Even more odd, his father invited Jim to lunch. The two went to a nearby café. "You can tell he's trying to treat this like a business situation because he has no real parenting skills," remembers Jim.

"So you've joined the army. What's your long-term plan?" Mr. Kennedy asked.

"I've joined the Green Berets, the Special Forces. It's the most advanced and highest caliber military training in the world." Jim rambled excitedly about the skills he was going to learn, not so much because he believed his father would be interested but because Jim had totally and completely fallen in love with the army. He also knew that the idea of it would scare his father: "Now he was at the end of my belt." It was the first time he'd ever excelled in a group environment in his life. He was bouncing in his seat.

Mr. Kennedy was surly and confused.

"And just what are you going to do with those skills?"

"I'm going to become a mercenary."

"What's that?"

Jim stared icily at his father. "People will pay me to travel around the world and kill people."

"I just delivered it as mean as I could, because I hated him," Jim says.

Mr. Kennedy turned white and looked as if he would slide under the table. "Kill people for a living?"

"After my training," continued Jim coldly, "I can get hired by anybody. Someone's going to want someone dead."

"I meant it," Jim says now. "It was my career path, and I felt completely qualified."

Flush with cash after spending six months accumulating paychecks, he headed for the mall to do some Christmas shopping so he could "dispense some charity on my siblings. I'm walking through the mall, and realizing I'm not going to make it through to the other end." Without the constant fuel of adrenaline that he had during basic training, he was in so much pain with shin splints and tendonitis that he couldn't walk. He waited an hour, sitting paralyzed on a mall bench before he could get up and slowly make his way out of the building.

When he returned to Camp Mackall in North Carolina, it was to begin Special Forces training. The passing rates for Special Forces were so low, and the training was so hard, that Jim, still in constant pain while standing, walking, and worse yet, running, was becoming concerned. And for the first time ever, he was encountering people who seemed as crazy as he was. "So I'm upping my game," he says. "Three weeks into it, I'm like, 'This is so good. Look at these people around me. They can barely take this.'" The worse the pain got, the further his physical limits were pushed, the better his performance became. In the beginning, soldiers dropped out quietly. No one talked about it. No one acknowledged it. One recruit went into a coma and was taken out on a helicopter. Watching his fellow soldiers fail or give up was strangely motivating. "It was almost like, the more they suffered, the more pain, the more agony . . . when they thought they personally couldn't do any more, my confidence just went up." When things reached torturous levels, "I'm brimming with smiles and humming marching tunes." Jim's cheery demeanor had the effect of scaring the shit out of everyone around him. "I sounded like a lunatic."

The tendonitis and shin splints got worse. While doing a field exercise, Jim's commanding officers recommended he drop out temporarily to recover, promising to let him back in the program afterward. "Usually they pull you out, you're done. But they could see I wasn't normal. They told me, 'Just for you. You'll not lose your place. You'll stay in.'" There was no way he was going to leave. He asked for something for the pain and was given a bottle of 800-milligram Motrin tablets, with no instructions on how to take them. "I was eating huge quantities. At one point, I would finish a bottle in a day." Throughout his training, he lived off handfuls of ibuprofen. "The reality was, I was poisoning my liver." It didn't matter. He would do whatever it took. He was in heaven. He had to stay there.

Although he loved the physical challenge, the gross ineptitude of the military organization was not lost on him. Things were often senseless and random, just like his father's rages. After an extended field exercise where it rained every day, and the men had to cross rivers and swamps, "All the sudden they go, 'Everybody, boots off! Field doctor's here!' The camp doctor comes and looks at our rotting feet and says, 'You need to wear dry socks and put foot powder on that.' Then he would go to the next guy and say exactly the same thing. And every guy is just laughing because everybody knows there is no foot powder and there are no dry socks, and we're in the field for another week. So this little vignette is so true in the military; they stare a problem down, state a solution that doesn't exist, and move on in an orderly fashion." It was an approach that one day Jim would see repeated over and over again, all the way up the ranks of the military, throughout the Department of Defense, and within the uppermost levels of the Pentagon.

On that same field exercise, Jim was in so much pain his team told him he was to play the role of the sniper victim. "Even though the pain was unbelievably excruciating, I couldn't stop on my own accord. So they made me." He feigned getting shot and pretended that he couldn't move. "And then the humiliating part happened. They made field stretchers, and they carried me. I was like, 'Fuck this. I'm a fucking burden.' They told me to shut up and enjoy the ride." Still, he could hardly appreciate it. "I was overwhelmed with guilt. I got a one-hour ride that day but was right back to it the next day."

At the end of Special Forces training, the final test of endurance was a course that lasted from sunup to sundown. "They drop you in, and you have a shitty map and a compass and you have to hit specific markers no bigger than a single iron fence post. You are not allowed to go in a straight line because that's how you get killed. And if you weren't back by sundown, you would be cut from the Special Forces." Jim took the map and the coordinates from the commander in complete confidence—even though when he walked, his swollen tendons rubbed on the back of his boots, causing agonizing pain. So when no one was looking, he removed his boots and took off. "Not only was I barefoot, I was so fast at it that I was coming back from targets and ran into others who were disoriented and lost. By the third one, I saw men panicking because they're not going to pass. And they're like, 'Can you help me?'" The men asking for help were officers who Jim had a lot of respect for, and he didn't want them to fail. He grabbed one guy, helped him finish the course, and then went back for another, finally finishing the course himself in the top twenty. He went back to where he'd hid his boots, put them on, and was cooking his C-rations when the commander came to the group.

"It's the strangest thing," the commander said suspiciously. "Everyone's reporting there are tracks . . . that someone was barefoot out there. What do you know about this, Kennedy?"

While field training was challenging for Jim only because of his medical issue, Phase 2 was a breeze. They gave the recruits a weapon that they were told to take apart and put back together. "It was just like in the cheesy movies," he says. He was able to beat the instructors' times. Somehow, his competence at all things military felt so natural. "I'm actually euphoric and feeling so unbelievably good. I'm becoming competitive, and I've never been competitive in my life before." Suddenly he cared about succeeding. Suddenly he wanted to be the best. Success became a drug coursing through his veins, filling up deep scars gouged in his childhood. Winning became so crucial that his newfound ego would drive him to make what he calls the single most humiliating, embarrassing, and devastating mistake in his entire life.

It was the final written exam in Special Forces classroom training. Miraculously, Jim received the second highest score of all the people in his class, missing only two questions. "I'd never scored well on anything before in my life," he says. The instructor handed back the tests and then began to read the correct answers out loud. He hesitated for a second, but being second place was no longer good enough. "What do you think I did? I changed two answers. In my mind, I'm like, 'You know what? They're telling us the right answers. I'll just mark the right answer.'" He turned the test back in, knowing they were to be regraded. This time he would have the highest score.

The story is still difficult for him to tell. "Think about my life," he says. "I was never measured to anything. Cheating was survival. The reason I was doing well in the army is that I had rules. This was explicitly breaking a covenant."

The trainers called Jim in.

"Do you want to explain anything?" they asked.

"No."

"Are you good?"

"Fine."

"Did you change the answers on the test?

"Well, yeah. You gave me the right answer."

"But you let us regrade the paper," the instructor told him. "This is unacceptable, and you know it. And so you are being ejected from the program."

Out of a million dark memories that Jim Kennedy kept in his mind's library, this was the one that would haunt him far into adulthood. "That mistake actually invalidated everything I achieved. I had the second highest grade of everyone I was competing against, most of them on their second or third enlistment, and I was kicked out by my own stupidity, a fledgling ego, avarice."

The sergeant major of the Special Forces saw Jim (who was waiting to be reassigned) at the gym. "'What are you doing here?' he asked. I said, 'I'm out. I'm done.' The next time he saw me there, he said, 'I'm getting you back in.' So he pulled every string, but couldn't get me back in." In the end, Jim Kennedy and all of his antics were too much. "The officer in charge didn't want me back in. So I got flushed. My sentence to purgatory was the Eighty-Second Airborne."

Trouble followed Jim to Fayetteville, North Carolina, where he began training with the red berets. As insolent as ever, he immediately ignored a basewide ban on visits to an area of Asian bars and clubs known for violent encounters between soldiers and the Asian mafia, including the dismemberment of more than one soldier. "It was the wrong place to be, so that was our favorite destination." Inevitably, one late-night adventure ended with a street fight and a few stabbed and slashed solders, one of them Jim's best friend. As his friend lay on the ground bleeding, Jim went after the assailant, a giant Caucasian man holding a large knife. "I stared him down, and I took a step forward." The attacker lunged at Jim, his knife slicing a deep gash in Jim's face from his temple to his cheek. Jim felt his face. It was wet and cold. The gash went all the way to the bone. He took off running, chasing the massive figure several blocks until he disappeared into a store. Jim stood at the storefront, pulling on the locked door, bleeding on the sidewalk. Half his face hung open. The police showed up.

"Get in the ambulance," a cop yelled at him.

A crowd was gathering. "The police hated being in the area and feared the situation was getting out of control. But I wouldn't leave. It was like a movie scene," Jim says. "I told them, 'I'm not getting into the fucking ambulance until you arrest him. I'll bleed to death right here.'" It was clear the police did not want to make the arrest, but a group of angry soldiers was forming a mob, forcing the police to go after the assailant.

After the assailant was arrested, Jim got in the ambulance. In the emergency room of the local hospital, a dentist who was on call sewed up the laceration on his face with seventy-two stitches. Jim was happy with the results; he figured he looked much better than he would if he'd ended up in a military hospital. "The next day, I'm ordered to the JAG office—the legal and investigative branch of the army. They tell me the guy is an enforcer for the Asian mafia." The military lawyers opened a file folder and showed him multiple counts of the perpetrator's violent crimes. Then they told him that he was not allowed to leave the base under any circumstances, because the assailant was out on bail. Jim was furious. He was determined to see the guy punished, so he called the district attorney's office and demanded they send the guy to prison. The DA assured him that their office intended to prosecute to the full extent of the law. If he was no longer in Fayetteville at the time of the trial, they promised to send him a ticket to come and testify.

Not long afterward, as Jim's wound healed into a four-inch-long pink scar, his company began a battalion-wide exercise that was to last several days. The first day did not go well. Jim was assigned to Alpha Company. "Our lieutenant was incompetent," he says. "We opened fire with blanks, and Alpha Company wiped out Bravo Company. Less than an hour later, the survivors of Bravo crossed our line of fire as we were engaging 'the enemy.' We wiped out Bravo Company twice." Not long after, an entire company of friendlies marched single file into Alpha Company's line of fire. "Friendly fire was rampant." Platoons, even entire companies, failed to reach appointed destinations, provide backup, or make rendezvous. The next day, Alpha Company was supposed to lay an ambush. The lieutenant, after doing badly in the exercise, suggested that Jim construct and lead the maneuver. He had limited time, but he used the basics of what he learned in Special Forces. The ambush was a complete success. "A few days later, word came down to the platoon sergeant that I was to be presented a garrison award. I refused. I told them I would not accept what the guys in the platoon called a 'bolo badge'—basically a bullshit award. I thought it was stupid."

The same week, there was a scheduled inspection of the battalion's barracks. Jim's company had the honor of receiving the battalion commander, staff, and a visiting two-star general. The company commander's ass was on the line, Jim knew. And although his superiors tried, no one could figure out how to get him out of the way during the mandatory inspection. "Inevitably, the general and his entourage landed outside my room. When asked by the general what I thought of the recent exercise, I could have told him what they all wanted to hear, but I have always preferred the raw, unpleasant truth."

"It was a disaster," Jim said, before laying out in precise detail the multiple friendly fire incidences, and telling the general that it was all the result of "a complete lack of land navigation skills." Though he didn't immediately realize it, he says, "I'd just signed my death warrant."

A week or so later, there was a companywide order to pick up trash on the grounds. Instead of wandering around chatting like the others, Jim sat down to read a letter he'd just received from his girlfriend. A superior officer saw him, and wrote him up for disobeying a "direct order." Jim couldn't believe it. Despite all the infractions in his military career, he'd never been written up for disobeying an officer. He was given a Chapter 13 (Separation for Unsatisfactory Performance) discharge. "This was total bullshit. The JAG, the military legal folks, told me the Chapter 13 would not stand up if I challenged it. I told them I didn't want to challenge it. It was time to go."

A few months later the discharge was finalized. Jim packed a small knapsack, left the barracks, and went out to his illegally parked primer-black '68 Dodge Charger. He sat in the driver's seat and stared dead into the future. For the first time, he saw a Jim Kennedy that he respected. He could compete at the highest level the U.S. army could dish out. He could own his future. "So from that second, I just accepted I'd be a different person. *When I leave here, I'm going to quit doing all this stupid shit. I'm going to have total control over myself and my life*. That was the last day I had a death wish. The day I left the army, I just said, *I'm going to fix my life. Somehow, I'm going to succeed.*"

6

A Better Mousetrap

It is always difficult to say what tomorrow will bring. With hope and optimism we believe that the rare earths will play a much more important role in science and technology than in the past.

—KARL GSCHNEIDNER, AMES LABORATORY, 1966[1]

More than a decade before Missouri State geologist Cheryl Seeger found rare earth elements at Pea Ridge, a young research scientist named John Croat arrived at the General Motors Research Laboratory in Warren, Michigan. It was 1972, and twenty-nine-year-old Croat had a fresh diploma proudly stating his doctorate of metallurgy from Iowa State University. He arrived with his schoolteacher wife, a profound analytical nature, and a deep knowledge of the way that rare earth elements behave. Ten years later, he finalized a discovery that completely revolutionized technology for the modern age.

At the time Croat landed in Michigan, domestic oil production was on the decline, global supply was down, and oil prices were skyrocketing. In response, General Motors and other manufacturers began to look for ways to build more fuel-efficient cars. Because one of the main factors in fuel efficiency is weight, General Motors had their researchers looking for any conceivable way to make their cars lighter. They looked at making them smaller and more aerodynamic. They tried making a lighter aluminum engine block, which failed dismally in the '79 Chevy Vega. The accessory motors that drove everything from windshield wipers to rearview mirrors, in particular, added considerable weight. One car could have up to sixty different motors, each containing an electromagnet made of heavy iron or ceramic, surrounded by a copper wire coil that induced an electromagnetic field to operate the motor. "It was well known that if the magnets were smaller and lighter, a lot of other components of the car could be designed to be smaller and lighter as well," Croat says. "If the motor to raise the window could be smaller, the door itself could be thinner and lighter." General Motors was the largest automobile manufacturer in the world. The sky was the limit as far as funding for new inventions went, so GM searched the country for a team to invent a new magnet that would be much lighter than those in use since the mid-1800s.

John Croat was the perfect candidate. Growing up, Croat knew early on that he did not belong in rural Iowa. After his father died, when he was just eight years old, the responsibility of the entire dairy farm landed on his mother and her seven other children under fifteen. "We milked dairy cows and sold milk, and that's how we lived," Croat says. "Five A.M. every day . . . and then every afternoon . . . three hundred and sixty-five days a year." At the time, farmers opted out of any type of federal social security program, so there was no government support for the Croat family. "The only thing that really saved us was the milking operation, which brought in a steady income." When Croat expressed an interest in attending college early on, "My mother wasn't at all surprised. I don't think she liked working on the farm either."

Croat's dream was to become an engineer, but his only option was nearby Simpson College. "The tuition was very cheap," remembers Croat, but there was no engineering department. He chose chemistry because he figured it was the closest thing. Croat's highly analytical mind took naturally to the science. Away from school, his life was far from easy. At night during the semester, he and his brother worked unloading freight from semi-trucks in nearby Des Moines. In the summertime, when school was out, he was back working the dairy farm, sunup to sundown. Summer was filled almost entirely by haymaking, laying up enough fodder to feed the cows through winter. "A lot of people looked forward to summer," he says. "Not me. I was always really happy to get back to school in the fall."

After earning his bachelor's degree, he headed to graduate school to avoid a routine technical job for the government or a large chemical company. Iowa State University in Ames was closely associated with the Ames Laboratory, a U.S. Department of Energy lab where some of the top researchers in the world were working on discoveries in the field of nuclear energy. While some of the work at the Ames Laboratory was aimed at nuclear reactor development and reprocessing nuclear fuel rods, most of the work was fundamental research focused on understanding of the physical nature of various materials. Another important mission of the laboratory was education, specifically the training of the nation's future research scientists.

While in graduate school, Croat needed to find a job to support himself. A position posted on the laboratory bulletin board offered work in the metallurgy department. The job also gave employees a schedule that would let them take off work to attend graduate level classes. "I thought it sounded wonderful. So I became a metallurgist," he says.

At the time, Iowa State was known among scientists as the rare earth Mecca, says Croat, and was considered to be on the leading edge of understanding the complex metals and minerals containing rare earth elements. Frank Spedding, a Manhattan Project team member, was the first person to invent—as an outgrowth of his work isolating uranium metal for use in the first nuclear reactor—a process to separate rare earths from the minerals that contain them, and then convert them into high-purity metals and compounds. The seventeen rare earth metals have similar chemical properties because their outermost electrons have the same configuration. The rare earths are also found intimately mixed together in nature, and because they have similar chemical properties, separation was found to be impossible using conventional metallurgical processes.

Spedding created what he called the "lanthanide contraction" process. Rare earth atoms get smaller as they get heavier. Thus the lower atomic numbers actually have a larger atomic radius, and then the radius "contracts"—by a subatomic fraction—as they get heavier across the periodic table. Using this infinitesimal difference, Spedding used an organic compound that would swap out atoms with the smallest radius first, then the next smallest. The technology was similar to a modern-day water softener. It was extremely slow and complicated, but what came out at the end was 99.999 percent pure. "The material that we had to work with at Iowa State was the predominant source of high purity rare earths in the world," Croat says.

It was at the Ames lab that Croat would not only learn to separate but become intimately familiar with the lanthanide suite of elements. He chose a Ph.D. project, with Spedding as his adviser, on the magnetic properties of two rare earth elements, lanthanum and lutetium, as well as two elements with similar properties that were often found in the same ore bodies, yttrium and scandium. In doing so, he became one of a very small group of researchers in the world looking at the magnetic properties of rare earths. There was only one problem: what to do with his obscure catalogue of knowledge. "I was really afraid that I would not be able to get a research job in the area of rare earths, because the field was so small. Also, when I graduated with a Ph.D. in 1972, there was a recession, and most new graduates were having trouble getting jobs." Croat nearly consigned himself to a "postdoc," a research position many took after a doctoral degree to continue working, usually in the same field, under an established researcher. However, the pay was paltry, and after spending several harrowing years on his dissertation, Croat was reluctant to delay starting a career. Then came a stroke of luck. "All of a sudden, someone called me up on the phone from the physics department of the General Motors Research Laboratories and asked, 'Are you interested in a job?'"

Croat's background and research experience seemed a perfect match for GM. They were aware of a newly developed magnet using the rare earth element samarium combined with cobalt metal. The magnet, invented by Karl Strnat at the Air Force Materials Laboratory in 1967, was so strong and so stable that even at very high temperatures, it would retain its magnetism. It was a "permanent" magnet, similar to a ceramic or ferrite (iron compound) magnet, but had much stronger magnetic properties. It also weighed much less than the electromagnetic magnets of the time. It was a promising invention, but samarium and cobalt were so expensive, there was no way they could ever be used commercially in automobiles. Croat's job, he was told, would be to find a way to make a similar magnet but with much cheaper ingredients.

Croat looked forward to his move to Warren, Michigan, and the magnet project with complete confidence. He and his wife, Kay, adapted well to suburban life. Kay was hired by the Fraser, Michigan, school system, and Croat, briefcase in hand, wearing a suit and tie to work for the first time in his life, headed to the massive modern complex that was General Motors Research Laboratories. Croat loved the laboratory, landscaped with beautiful trees and a large manmade pond. "It was a wonderful place to work," he says. The best part: it was full of science and discovery, "just like a college campus."

Croat immediately dove into the problem they'd hired him to solve: develop a lower-cost, high-performance magnet with properties similar to samarium-cobalt. "When we started looking at samarium and cobalt, it soon became obvious they were never going to be cheap enough for our applications in cars." There was a civil war in the Belgian Congo, and the price of cobalt went through the roof, to around $60 a pound. "Since the discovery of samarium-cobalt, scientists everywhere had been asking, Why can't we make a rare earth magnet with iron?" remembers Croat. He liked the idea, so he tried to replace the expensive samarium with one of the cheaper, "light" rare earths (those with lower numbers, moving to the left on the periodic table) like cerium, neodymium, or praseodymium, and also replace expensive cobalt with super cheap iron. However, he "soon found that the magnetic properties dropped faster than the cost—always a losing proposition."

Croat thought and thought. He knew that a material's "coercive" force—its resistance to losing its magnetism—comes from the way electrons spin around the nucleus of an atom. He also knew that, of the light rare earths, only neodymium and praseodymium had the proper electronic structure to hold a magnetic force. "However, there was a fundamental and seemingly intractable problem," he remembers. While samarium and cobalt combined to make an "intermetallic" compound because of the way their atoms interlocked, neodymium or praseodymium atoms did not fit with iron atoms the same way. And there was no way to form a magnet without an intermetallic compound. Croat's team, which consisted of three others—all physicists—had yet to get their hands dirty in the laboratory. "We never did any actual experiments trying to make magnets from these materials, because we knew that nothing would work," he says. "We just thought about it, because that was an insurmountable problem."

Fortunately, a lightbulb went off for Croat in 1975, in the form of a scientific paper by Arthur Clark, a research scientist at the Naval

Ordinance Laboratory.[2] Clark discovered that heating and then quickly cooling a mixture of the rare earth element terbium, mixed with iron, gave a very strong magnetic force. The magnetism did not hold up over time, but it was there, and it was very interesting to Croat and two of his fellow coworkers, Jan Herbst and John Keem. The team thought that they could try the same thing with neodymium and iron, and praseodymium and iron.

Over several years Croat tried a myriad of experiments using each of the light rare earths—or a mixture of them, combined with iron. His ideas were limitless, and so was his budget. General Motors held 60 percent of the automobile market since the 1950s and was so flush with cash that the government was talking about breaking it up into its separate units: Chevrolet, Pontiac, Buick, Oldsmobile, and Cadillac.[3] "Within certain guidelines, we were allowed to do pretty much anything we wanted to do," Croat says. One new process he wanted to try was called melt spinning, an invention from the Idaho National Lab, in which a molten stream of alloy is directed onto the surface of a cold, rotating disk to produce a rapidly quenched metal ribbon. He hoped melt spinning would cause the alloy to become highly magnetic. But despite repeated attempts, all of his samples developed disappointingly small levels of coercive force.

The years churned by, and Croat was called to work on several other projects. Whenever time allowed, he would return to his hypothesis that rapidly solidified iron combined with neodymium or praseodymium could somehow be made into a permanent magnet. Then, in early 1982, Croat added the element boron to his mixture. "That gave an immediate and profound effect, more than doubling the magnetic properties I'd achieved previously." At first, it seemed too good to be true. After so many disappointments over so many years, Croat was sure his new mixture would lose its magnetism when heated. It didn't. It was the determined scientist's Eureka moment. John Croat, former Iowa farm kid, had created the world's first commercially viable superstrong permanent magnet using neodymium, iron, and boron.

Or so he thought.

In a laboratory halfway around the world, a lone Japanese scientist named Masato Sagawa was putting the finishing touches on his own version of the neodymium-iron-boron magnet. Sagawa, like Croat, had come into the world of magnetism by chance. Sagawa was born in 1943 in Tokushima Prefecture to a father who entertained his three children with stories of new scientific discoveries. "I remember that at six years old, I was very excited about an article in a newspaper which reported that Professor Hideki Yukawa won a Nobel Prize in Physics. It was the first time in Japan," Sagawa says.

It was the early 1960s when Sagawa entered college, and the young man saw the writing on the wall of Japan's technology revolution: "I realized that engineers were playing the most important parts in the growth of the economy." So he went to school to become an engineer. While he excelled analytically, he found he lacked creativity and had a hard time coming up with his own ideas for a Ph.D. thesis. He eventually eked out his doctorate at Tohoku University, but no major scientific journal would publish his work, and in the publish-or-perish university culture, he was left squarely in the "perish" pile. Unable to pursue research at another university, Sagawa landed a job at Fujitsu Laboratories in 1972 as part of a team of materials scientists looking at ways to make magnets for better electric switches. It was Sagawa's first look at magnetism. "I was not confident with the subject at first, but became more interested as I carried out the research," he says. He also realized that he was actually an excellent researcher when given a specific project, "Something that tends to be more clear in companies than in universities."

A few years later Sagawa was given new orders: Develop a new samarium-cobalt magnet with improved mechanical properties for newly developed switches. He took the appointment as an honor, and as a Japanese scientist, a natural destiny. Japan had always been at the forefront of finding ways to produce permanent magnets using a mixture of metals. In 1916 Kotaro Honda, one of the most renowned physicists in the world of magnetism, began mixing cobalt with steel to increase the magnetic properties of the steel. In 1931 metallurgist Tokushichi Mishima added aluminum to nonmagnetic steel, creating Alnico permanent magnets, an important invention with practical uses in electric motors, loudspeakers, and various sensors.[4] And in 1976 Yoshio Tawara tweaked the recipe for samarium-cobalt magnets to produce the strongest permanent magnet yet.[5] "I felt that it was normal for me to develop a new permanent magnet with the highest magnetic energy product as a researcher in Japan," Sagawa says.

While cobalt showed a lot of promise in making permanent magnets, Sagawa noted that many scientists around the world were only concentrating on rare earth magnets combined with cobalt. Like Croat, Sagawa turned to what seemed obvious: the most plentiful metal on earth— which also happens to be the most magnetic. "Iron is much more abundant than cobalt, and moreover, iron has a much larger magnetic moment," he says. Unfortunately, Fujitsu did not give Sagawa the free rein that Croat had at GM, so he had no means to pursue his brainchild.

Still, ideas about magnetism percolated in Sagawa's mind, invading his dreams. "I wanted to develop a rare earth iron permanent magnet, but I had no idea how to realize it," he says. In 1978 he attended a symposium where a researcher named Masaaki Hamano offered a simple explanation as to why the rare earth and iron compounds couldn't make practical permanent magnets; the spaces between iron atoms that formed magnetic crystals were too narrow—like a dumbbell with a handle too short to grip with your fist. *Aha!* thought Sagawa, who immediately tried out an experiment in his head. He could take carbon or boron, the two elements with the smallest atomic diameter, and try to force them in between the iron atoms and expand the spaces in the crystal. Sagawa went back to his lab at Fujitsu and played around with his ideas, mixing iron and rare earth elements with carbon and boron. He was able to make stable compounds, but not permanent magnets. Still, he knew he was onto something.

Shortly after, Fujitsu made a companywide decision to give up on research involving permanent magnets. The world was changing. Mechanical switches—made with magnets and conducting metals—were being replaced by semiconductors. Fujitsu was a switch company and had no need for Sagawa's magnet ideas. Sagawa resigned himself. "It was a normal decision, and I had to follow it," he says. His bosses put him on other projects, but he wasn't happy with his new assignments. Every weekday at five o'clock, when the workday at Fujitsu ended, Sagawa's mind returned to his rare earth magnets. His "thinking experiments" consumed him. Without asking permission from his bosses, he came to the empty lab on his holidays. His wife, Hisako, completely supported his desire to work away his free time, he says. Alone with his metal powders and beakers and machines, he brought the ideas in his head to life, working to confirm or discount his hypotheses.

In early 1982, even after his promotion to chief researcher, Sagawa was miserable at Fujitsu. In the days of wild economic growth in Japan, corporate managers had the reputation of being very hard on their employees. After a severe scolding by his boss, Sagawa couldn't take it anymore. He submitted his resignation the next day, but his boss demanded he stay another three months and refused to give him his earned vacation. His heart sank. He was desperate to get out of Fujitsu and work on his true passion. "So I asked him to allow me to work on experiments for my next job after hours, and he agreed."

In the empty magnet laboratory, long into the evening, Sagawa made many experiments and tested new ideas. Then one day, all the dreams and thinking experiments of Masato Sagawa resulted in a brilliant and world-changing reality. He used high pressure and temperature, in a process he called "sintering," to form a little bar of neodymium, iron, and boron into a superstrong permanent magnet. "I leaped up with joy so high, I almost reached the ceiling," Sagawa says.

Soon after, he went to interview with a company called Sumitomo. After telling them of his passion for rare earth/iron magnets, the company president, Norishige Okada, offered Sagawa a deal almost too good to be true. "Mr. Okada said that I would have indefinite funding and as many engineers as I needed for my research," says Sagawa. Okada and the entire Sumitomo company could not have imagined that their investment would result in so quick and so brilliant a return. Three months after arriving, Masato Sagawa tripled the strength of his sintered magnet—creating the strongest permanent magnet in the world.

Back in Michigan, Croat continued perfecting his magnet while lawyers at General Motors put together the paperwork to patent his discovery. Because patent applications were open and public in those days, GM wanted to wait as long as possible to file the patent so that they could keep Croat's amazing discovery under wraps. Croat was also forbidden to publish or talk about the results of his work until the patent was secured. The process was painfully slow. "The basic problem was that the paperwork kept getting changed for one reason or another," Croat says. "There were no computer word processors at that time, so the patent had to be retyped again and again. This seemed to take an interminable amount of time, and months were lost."

So it wasn't until the GM attorneys finally went to file the patent that they found that one Masato Sagawa from the Sumitomo Corporation of Japan had beaten them to the patent office by thirteen days. Croat's heart sank. "That was very disappointing when I found out they had filed in the United States before us," he remembers. However, at that time, the U.S. Patent Office operated on a first-to-invent system rather than a first-to-file system.[6] Because Croat could document proof of his invention back to February 1982, he ended up with all the U.S. patents, while Sagawa got the patents in the rest of the world.

Sagawa was not privy to the patent battle. He didn't find out about the American magnet inventor until later that year, at a conference in Philadelphia. Looking at the lineup of talks, he felt a knot in his stomach. "I was surprised to see the conference program, and that several American groups were giving presentations on the same neodymium-iron-boron permanent magnets," he says. "But I was relieved that the process to produce the permanent magnets was much different." Although Sagawa thought that his sintering process was superior to Croat's quenched method, he had no proprietary feelings over the discovery of the magnet mixture. "I thought that the American groups represented by Dr. John Croat and my group, represented by myself, should share the invention fairly."

"Whenever I say anything, I say Sagawa and I invented neodymium-iron-boron, simultaneously," says Croat, who also gives credit to many of the researchers whose discoveries helped him and Sagawa reach their goals. "A famous man said he could see so far because he was standing on the shoulders of giants," says Croat. "This is certainly true here. Both Sagawa and I are greatly indebted to scientists like Spedding, who first produced pure rare earths, and others who studied intermetallic compounds. Without the basic framework, our inventions would not have been possible."

With the patent issues resolved, General Motors allowed Croat to go public with his invention. Croat—who, at six foot two inches tall, towered over his colleagues, looking more like a smooth-talking salesman than a scientist—and his GM team attended many conferences to talk about the discovery and science of his magnets. News of the fascinating invention soon reached nearby Anderson, Indiana, and pricked up the ears of Joe Lehman, an engineer turned manager at the Delco Remy Corporation. A division of GM, Delco Remy produced coils, distributors, and ignition switches. But the product that most interested Lehman was the starter motor, a big, weighty motor made super heavy by its massive induction magnet, which needed to produce a force strong enough to kick over a car's gasoline motor.

The cranking motor in production at Delco Remy in 1983 was supplied not only to GM for their cars but also to many of the other auto manufacturers in Detroit. It was designed in the early twentieth century by Charles F. Kettering, a young engineer who ran a start-up electronics firm out of a barn in Dayton, Ohio.[7] It all began when Kettering serendipitously crossed paths with Henry Leland, the president of Cadillac Motors, in 1910.

As it turned out, a dear friend of Leland's had died in 1908 from injuries sustained while attempting to start a car with a hand crank. Leland promised Kettering that if the "barn crew," as Kettering and his former colleagues from the National Cash Register Company called themselves, could build an electric starter motor, Leland would install it in his Cadillacs. Kettering's first attempt was massive, unbelievably heavy, and clearly unworkable on an automobile. They redesigned it. Finally, on Christmas Eve 1911, the very first electric starter was installed in a 1912 Cadillac. The barn crew became Dayton Engineering Laboratories Company (Delco), and Delco cranking motors turned over the motors of every car coming out of Detroit. It was an automotive revolution as the "cantankerous hand-crank" became obsolete, and automobiles revved to life with the push of a pedal and eventually the turn of a key. One Detroit newspaper wrote that Kettering "did more to emancipate women than Susan B. Anthony."[8]

Seventy years later, Joe Lehman was ready to revolutionize Kettering's starter motor. He dreamed it could be smaller and more efficient. In the mid-1970s, Delco Remy tried samarium-cobalt magnets in motors, but like everyone else, ran aground on their cost. Lehman tried having his engineers build a new version with a permanent iron magnet for the Corvette, but the magnet wasn't powerful enough, and the design never made it off the ground. So when Lehman heard of Croat's magnet, he was licking his chops. "It was a magnet that had about double the energy product range of conventional ferrites (iron magnets)," Lehman recalled in a 1985 Delco Remy newsletter interview.[9] "We didn't know exactly how it was made or what the technology was behind it—we only saw charts—but we were interested."

Lehman went personally to the GM laboratories and told them all about his idea for a new cranking motor. "And we said if it were possible to make an economically feasible boron neodymium and iron magnet for it, we needed to do it," Lehman said. "We knew Delco Remy couldn't do it alone because of the high level of technology involved in the totally new processes that would have to be developed." Croat was delighted by the immediate interest in his magnet, but he really had no idea how to take his invention from the lab to the assembly line. Up until that time, the most material that had been produced at one time was ten or twenty grams. "The question was how to scale up the process, and how to build production melt spinners."

Once the development was approved within Delco Remy, Lehman contracted a California company to build a much larger melt spinner. The machine was built quickly and set up at Delco Remy. "However, we were having trouble, lots of trouble, getting it to run," he said. The next stumbling block was that they were using material that had never been used and was nearly impossible to find. "All I had done was use just a small amount of neodymium in a laboratory melt spinner. We needed hundreds of pounds. Purchasing at General Motors told Croat, "All the neodymium in the world, you could put in one suitcase." As for praseodymium, there was even less, as the metal was a third less abundant in rare earth ores. "So we had to develop a supply system." Fred Brewer, Delco's chief chemist of process engineering, was put on the job. He found 80 percent of the neodymium they needed from a French company called Rhône-Poulenc that was importing ore from Australia and separating it in La Rochelle, France, and another 15 percent from a company in California called Molycorp that had a mine in the mountains near the Nevada border. Molycorp had never processed neodymium before, so they sent rare earth oxide to the East Coast for processing. The final 5 percent, Brewer bought from a Chinese company called Yu Long.

Joe Lehman, smoking like a chimney and flustered by his lack of progress in getting the melt spinner to work, convinced Croat to come to Delco Remy permanently. Croat and his wife packed up and moved to Anderson, Indiana, and "not too long afterward, the thing started working." Croat says that he doesn't take the credit. "I attribute a lot to Joe Lehman, who was the driving force behind the pilot development." Lehman, thrilled with his success, decided the magnets alone were worthy of an entire new business unit of Delco Remy. The new division would be named after the way the magnets were made by instantaneous quenching: Magnequench.

"We started to get excited because we recognized that we could evolve into the world's best permanent magnet," Lehman said. "It had taken thirty years for ferrite magnets to go from an energy product of two to an energy product of four. Here we had gone from seven to fifteen in a few months."[10] (The energy product of a magnet is the amount of magnetic energy it can store, measured in kilojoules per cubic meter.) Bob Van Wingerden, the new Magnequench manager, said in a Delco Remy newsletter interview, "It's the old cliché: invent a better mousetrap and the world will beat a path to your door. John Croat has literally given us that better mousetrap." Van Wingerden had high hopes for Magnequench: "I'd say that in 10 years, more than 80 percent of our sales will be outside of GM. Furthermore, I'm sure we'll be in markets we haven't dreamed of today."

Lehman organized a press conference at the local union hall to announce Delco Remy's new $70 million investment in a Magnequench manufacturing plant. Union representatives, the mayor of Anderson, and even the lieutenant governor arrived to hear about the new technology company and test the strength of Croat's magnets. Reporters came from around the state. A news helicopter landed on a nearby field. As the press watched, and with bit of fanfare not really in his nature, Croat held a metal plate while the lieutenant governor struggled unsuccessfully to dislodge a Magnequench magnet.

"I'm pleased to announced that Delco Remy is bringing exciting new technology to Anderson," Magnequench's general manager, Jim Ault, told the gathered crowd. "This is significant to central Indiana. Now we too can say we are the home of high tech in a world-class climate." Union representatives hailed the hoped-for new jobs and promised to build a mutually beneficial partnership with the company and the community.

"I knew it was a big scientific invention, but I didn't realize how commercial it would become," says Croat, who said a few words to the gathered press that day. "We had no idea how fast the market would grow."

Sadly for Joe Lehman, when Delco Remy engineers finally got around to building and testing the cranking motor, his dreams were dashed. The entire mechanism for which Lehman had brought the magnets to Indiana crumbled under the natural influence of the environment. "The problem was, we didn't understand how much protection the magnet needed from corrosion," says Croat. The engineers soon realized that the magnets broke down if subjected to moisture or salt. "And of course," Croat says, "a cranking motor is the worst environment in the world."

After the test motors began to fail, the program was shelved. "Lehman was very disappointed, because that was his baby," says Croat. "We all were disappointed." By the time Croat and his team found ways to protect the magnet against corrosion, it was too late. General Motors was running out of money for extensive new-product research. The 1986 Corvettes and Buick Somersets that were supposed to have new Delco starter motors rolled off the assembly line with the motors invented by Charles Kettering in 1911.

Magnequench's failure to immediately revolutionize automotive production did nothing to dampen the scientific community's excitement, though. In 1986, both Croat and Sagawa received the James C. McGroddy Prize for New Materials from the American Physical Society "for their pioneering research on the preparation and characterization of rare earth-iron-boron materials which led to the discovery of a new class of permanent magnets of unusual scientific interest and technological promise." At the ceremony's reception in Las Vegas, Croat and Sagawa met for the first time and developed a friendly rapport.

The American Physical Society couldn't have been more prophetic. That year, Jeff Day, the marketing manager of Magnequench, took several samples of the melt-spun powder, a few magnets, and some technical data to a number of Japanese motor manufacturers, including Panasonic and Matsushita. He returned to Indiana triumphant. Around the world, the personal computer was becoming a reality for the common consumer, and the industry had exploded with games, CD-ROMs, printers, and other electronics. Magnequench magnets were perfect for spindle motors used in hard disk drives, as well as motors for many other consumer electronic products. Companies that wanted to produce their own magnets convinced Magnequench to sell them the patented neodymium-iron-boron powder. In the early 1990s, sales grew rapidly. Sagawa's sintered magnets became even more prolific as the Sumitomo Corporation poured all of the company's resources into the brilliant invention. The sintered version worked incredibly well for input and output devices on computers. Headphones and speakers became smaller and more streamlined. Sumitomo licensed the technology to companies across the world, and each year, the production of new magnets increased 200 percent. Inventors designed new products using the supermagnets, and engineers souped up old electronics. The miraculous invention of rare earth permanent magnets ushered in a technological revolution at breakneck speed, with Japan and the United States leading the charge, compliments of Dr. John Croat and Dr. Masato Sagawa.

Over the following decade, the United States embarked on a technological revolution that seemed possible only in the realm of science fiction. The mastery of the young nation fueled the country's collective ego. Paradoxically, in Washington, D.C., a congress full of professional politicians, lawyers, and businessmen were quick to abandon the foundations of science that brought the country such unparalleled success. Newt Gingrich, a trained historian, became Speaker of the House in 1995 and soon after shut down the Office of Technology Assessment—the department that gave lawmakers advice on science and technology.[11] Now lawmakers could rely on whatever "industry experts" they chose rather than a nonpartisan scientific body. Gingrich then marched in renowned "science skeptics" to poke holes in well-accepted theories of ozone depletion and greenhouse gas, calling the junkets "scientific integrity" hearings. The Republicans ate up the industry-funded outliers' message; burning fossil fuels couldn't possibly be harming the planet. In 1996 Congress shuttered the U.S. Bureau of Mines.

Meanwhile, lawmakers preached Milton Friedman's free-market ideology. Not only would scientific research be beholden to the market but so would every other sector—healthcare, prisons, even America's national defense. And if American workers couldn't compete in the free market? No problem. Congress paved the way for companies to leave the United States and make products for America in China. Wall Street boomed. America was on top of the world, propped up by previous decades of sound economics and government spending on research and development. No one was paying attention while, with incredible skill and precision, China began to dismantle the intellectual foundation of American companies that did business on the Chinese mainland. Few in America saw a downside. There wasn't much thought given to competition by an ancient and closed civilization. All the while, Chinese Communist Party leaders—trained mainly as engineers and scientists—plodded along methodically, brilliantly planning and gathering intelligence—not through clandestine means, but with the warm invitation and legal consent of the United States government.

7

Homecoming

To his brothers and sisters, Jim Kennedy had always been different. When he came back from the army in 1985, he was a complete alien. The Kennedy siblings could not understand his new sense of direction and self-imposed order. "He acted like a staff sergeant," says his mother. To his sister Mary, he was intractable: "Everything to the new Jim Kennedy had to be black or white, right or wrong." He was very political—a staunch advocate of a conservative Republican platform. "When I finally decided my life mattered, it was all or nothing," Jim says. "There was no halfway."

One thing was certain. Jim was not going back to live in his father's house. With the money he saved while in the army and his new job as a machinist in a nearby factory, he rented a studio apartment above a dry-cleaning business. "He finally broke free," remembers Mark Haselhorst, who was still very close to Jim and thrilled he was back from the army. However, much to the dismay of his hard-partying friends, Jim had shed his wild ways and didn't drink anymore. Mark had no intention of quitting drugs or alcohol. "But the fact that Jim wanted to go forward with his life and sober up didn't diminish our friendship," he says.

The apartment was a complete dump. Jim, like all his siblings, was adept at fixing things. He asked Mark to help remodel the place. "It hadn't been lived in for years, and we were over there and turning it into a very nice first apartment; plastering water-damaged walls, painting and refinishing the floors," Mark says. As they were putting on the finishing touches, Mr. Kennedy dropped by. Mark always felt uncomfortable around Jim's father, and he could tell that the freshly remodeled apartment—which Jim had obviously worked very hard on—was making the senior Kennedy angry. Oddly, Mr. Kennedy asked both Jim and Mark to come to the Kennedy house for dinner. Later that night, when Mark pulled into the Kennedys' driveway, Mr. Kennedy approached the open driver's-side window of Mark's car. "I was seated behind the wheel. He looks in my lap where I have a beer between my legs. I see his eyes focus on the beer." Then, with all the force the aging boxer could marshal, Mr. Kennedy cold-cocked Mark in the face.

Mark was shocked, then furious. "I was going to tear him apart. *I'm twenty-one now, and your ass is mine. I'm gonna fucking kill you.* I got out of the car with such rage. I wasn't a phenomenal fighter, but I was going to go after him with a vengeance, and I was going to hurt him as bad as I physically could." Mark had seen Mr. Kennedy's violent behavior since he was a young teenager, but this was beyond anything he could believe.

"As I got out of the car, Mr. Kennedy was stepping back. I locked eyes with Jim for a moment, and there was some kind of weird thing where Jim conveyed to me: *If you hit him, he wins.* I backed down because Jim's friendship meant much more to me than hurting his dad." Jim was thankful Mark was a good enough friend to take that punch and walk away. Mark was young, tall, and strong, and he could have hurt Jim's father badly. Jim thought for a minute he might enjoy watching the person he hated most in the world get his due. But it wasn't the way he wanted it to be. It was not how the new Jim Kennedy wanted to live his life.

A parade of Jim's broken friends and several of his siblings came to live in his new apartment. "My rules were simple. You just have to do something positive, something constructive with your life." He never charged anybody rent. "My sister Kathy stayed with me. My brother John stayed with me." Mary remembers being pushed to run with him in the early morning while she lived there. "My apartment was a haven for family members to get away from our father. John, Jeff, Mary, and Kathy all lived with me. Their stays were short, as I required them to demonstrate life goals and direction," Jim says. Rarely did things end well. "Eventually they went back to the chaos of home because it was an expectation-free environment." Bad blood trailed them out the door. "He helped his family members in every possible way, and most of them eventually turned on him with utter hatred," says Mark.

Jim reunited with his former girlfriend, Mary Egan, who invited him to her parents' house for dinner. "To any outside observer, it was a comedic spectacle," he says. "Mary brought home a stray, and her parents—a doctor and nurse—were too polite to protest." Fearless as usual even in this completely foreign environment, he was determined to make the most of it. Rule one from his survival training was "Adapt to your environment." He watched how they put their knives and forks on their plates, their elbows resting below the table edge. He put a napkin in his lap, was always thoughtful of his posture. He made a point of following and taking part in their weirdly polite dinner banter. While he could blend in to a point, he lacked any cultured interests to add to the conversation. Still, he tried, always ungraceful but clearly sincere. "Her parents responded in a caring way. They even tried to help me and lend direction," he remembers.

Mary's father was a history buff, and Jim found he had a real interest in history as well. Dr. Egan lent him a book on the history of military logistical developments, from Philip of Macedonia through the Roman Empire. Jim painstakingly struggled through it. As he did, his mind began to open to the idea of a much bigger world. Soon, awkward dinner conversations transformed into thoughtful and well-rounded discussions. Jim was holding his own. Mary's older brother was smart and charming. He and Mary attended the best local prep schools, and were now excelling in college. Jim started to believe that learning in a structured environment could actually be a good thing.

While Jim was determined not to let the chaos of growing up with his father derail his mission to live a better life, there was something about Mr. Kennedy he deeply respected. Shortly before Jim joined the army, his father jumped ship from his highly successful career at Monsanto to form his own investment firm, Kennedy Capital Management. Along with his older brothers, Jim helped their father design the office layout and move furniture into the empty rooms. And although it pained him to admit it, he couldn't help but admire his father's decision to leave a twenty-five-year career to build his own business—showing a fearlessness that had not wavered since Gerald Kennedy was very young.

Gerald Kennedy's strange path from corporate management to Wall Street trader went like this: As a teenager in the 1950s, while still on the family farm in Iowa, he began investing in the stock market. And in doing so, he figured out how to make enough money to pay his way through college. He understood the business of Wall Street trading, of equity, of gains and losses. The mathematics of Wall Street, the analyses and projections, all made perfect sense. He had a brilliant scientific mind, and after earning his degree in chemical engineering from Iowa State University, he went to work for Monsanto. In his spare time, Mr. Kennedy interpreted the techniques of famous investors, including mutual fund pioneer John Templeton.

With a strong background in statistical analysis, Mr. Kennedy looked at Templeton's strategies and "reverse-engineered" Templeton's stock selection process; that is, he worked the data sets backward to learn common characteristics of investments with positive results. While still at Monsanto, he realized that his prospects for making big money were dim in a corporate career, and he pitched his theories in a letter to Templeton. Impressed by Mr. Kennedy's in-depth analysis of his investment style, Templeton dropped into St. Louis to meet this intriguing amateur investor. Shortly after their lunch in the Monsanto cafeteria, Templeton handed Mr. Kennedy a huge win, entrusting the corporate executive with one of his personal accounts. It was exactly what Mr. Kennedy needed to strike out on his own. Kennedy Capital Management of St. Louis began to grow. Mr. Kennedy's charm and charisma quickly won him clients from union pension funds to wealthy St. Louis business owners. "Kennedy Capital grew into that perfect American story of success," says Jim. "No one worked harder."

With his father in finance and without any other idea for a major, Jim signed up in the fall of 1985 for his first community college class: Introduction to Business. One of his first assignments was to read the *Wall Street Journal* every day. Determined to excel, Jim picked up a copy and began to read. It was the first time he'd stopped to analyze why he'd avoided reading his whole life. "It was brutally difficult. It took me an hour to read the front page." Sometimes he had to read a sentence two or three times. Jim's military instinct told him that his inability to read well was incompatible with surviving in a civilized environment. His training also taught him that if you have a weakness, make it your strength—so that's what he set out to do. Now he *forced* himself to read. What he found was that, as hard as it was for him to interpret the black lines and squiggles on a page, the vast area of untapped gray matter in his brain eventually absorbed the ideas that Jim now forced upon it. As he read, his mind processed and stored and churned. It was not easy. At first, every word would seek to deceive him. He continued to try, and eventually he learned that all the words before him could be compelled to behave. His slow and disciplined reading led to quick comprehension, and his new skill seemed to offer limitless possibilities.

It was only later that Jim first heard the term *dyslexic* in a sociology class. "The lights went on in a very dark room," he says. For the first time he started to understand a part of the reason for his personal failures. "When you're dyslexic, you avoid reading. When you must, your mind jumps ahead of the words on the page, trying to recognize clues, context, patterns, and then your brain assumes what the word is. It takes a bunch of guesses—and it's mentally exhausting." Jim says that the idea that a dyslexic person sees letters inverted or backward is wrong. "You would never even notice that. Your brain can't see the difference." Growing up, Jim worked around his dyslexia by ignoring any and all assignments in school. "I just never did any homework. I never did any papers or did the required reading."

Given a glimmer of this new world of unlimited knowledge, Jim became obsessed. "I would never listen to the radio. I would never watch TV, movies, or do anything that wasn't actually productive. If I was awake, I was reading and studying. If I was driving in a car, I was listening to books on tape." He read the classics. He read politics and economics. His favorite place in the world was his leather reclining reading chair. He joined the Book of the Month Club, Eaton Press, and the History Book Club. His mailbox exploded with leather-bound classics, new nonfiction, and hard-cover histories. He read every chance he could. A few times, he tried reading in the shower, holding the book outside the curtain while soaping up. Armed with a new intellectual confidence, he signed up for a full load of classes at the community college.

But before the semester started, there was some unfinished business Jim needed to resolve. The scar on his face had healed, but it would never go away, and every time he looked in the mirror he felt the burn of injustice. There was a criminal walking the streets who'd stabbed him and his friend.

Jim needed to make sure that his attacker was put away for a very long time. Unfortunately, the Cumberland County district attorney's office in North Carolina still hadn't called him to testify. Jim kept calling and checking for the trial to appear on the docket in the Fayetteville courthouse, so he knew when the date was set. But no one at the prosecutor's office would return his calls. "So I bought my own ticket and flew in," he says. And in doing so, he would start a lifelong pattern of never walking away from an issue until the resolution was exactly where he wanted it to be.

Sitting in the courtroom, Jim couldn't have looked more out of place. Fayetteville was in a poor rural county in North Carolina. Most of the people in the courtroom looked like hillbillies to Jim, and he wondered about the potentially sordid backgrounds of the jury pool. Most were unemployed. Some were unskilled laborers. A few actually worked for the family of Jim's assailant. By contrast, Jim—who still looked very young for his age—was wearing his new wire-framed glasses, a three-piece designer suit, and a Seiko watch he'd purchased while in the army. "My dad did one miraculous thing for me in my life," he says. "When I graduated from high school, he took me to one of those discount suit warehouses. The place was full of off-style suits and oddball sizes. Having a twenty-eight-inch waist and a forty-inch chest, I fit nicely into the oddball size category and hit pay dirt." The salesman was clearly excited to unload a one-off Geoffrey Beene suit with such strange dimensions. Mr. Kennedy bargained hard for a better price and purchased the suit for "something like ninety dollars," says Jim.

Jim watched in disgust as two people who were directly related to the defendant were allowed to sit on the jury. The defense attorney and most of the other lawyers present were unshaven. Their shirts were grimy at the sleeve and collar. Even with Jim's background, he was overwhelmed with the backwardness of it all. When the trial began, the defense attorney saw Jim and immediately chose a tactic. "Look at him in his fancy suit and wearing a fancy watch he bought with his daddy's money—coming down here and startin' fights," the attorney prodded the jury, pointing derisively at Jim. "The jurors are looking at me," he says, "and thinking, 'Look at you with all your money.' I might as well have been one of those liberal Massachusetts Kennedys." He was seething, but told himself that justice would be done when he had his turn on the stand. He'd tell his eyewitness story of the attack by the defendant, who now sat glaring at him from the defense table. When it seemed like it was time for him to be called to testify, though, the prosecutor told the judge that he had no questions and no witnesses he would like to call. Then the judge called a recess.

Jim was livid. But as usual, faced with challenging circumstances, he instantly came up with a plan. *They think I'm someone important in this suit,* he realized. He followed the lawyers into a private area for court employees and attorneys, and found the prosecutor and defense attorney sharing a laugh. Jim moved deep into the prosecutor's personal space, speaking loud enough for everyone in the small room to hear him: "Listen to me right fucking now. If you do not put me on the stand after this recess, I am going to make one phone call. My father is going to send his team of corporate attorneys down to Fayetteville on the corporate jet, and we are going to take this case to the State Supreme Court." It was a total bluff. Jim was not on speaking terms with his father. The company did not have any corporate attorneys, or a jet. In fact, Kennedy Capital Management was still struggling to get off the ground.

A few minutes later, the prosecutor called Jim to testify, "clearly motivated to avoid a confrontation with any out-of-town lawyers who could expose their small town political corruption." It was pure fiction, but it played out perfectly. The trial ended in a guilty verdict. Jim's assailant got ten years in prison.

8

The Student

Back in St. Louis, Jim started school with a vengeance. He worked full-time in the machine shop, took construction jobs on weekends, and carried eighteen credit hours. In his free time, he read the entire works of Ayn Rand, Aldous Huxley, and George Orwell. "I would read and study until I would collapse of exhaustion. My last semester at community college, I got mono, and I was jaundiced. My apartment stank like someone was dying. My friends would come over and say, 'What the fuck is wrong with you?'" From the second he woke up until the second he went to bed, there was nothing in Jim Kennedy's life for relaxation or pleasure. "I had to replace twenty years of zero discipline and indifference in just four years of college," he says. In 1988, after two years, Jim graduated from community college with a 3.9 GPA. His confidence was high, and he planned his next move. At the family Christmas party, his father asked him what his plans were with his "big associate's degree." His oldest brother, Jerry Jr., a recent graduate from a technical college, was playing sidekick to Mr. Kennedy's taunting.

"I'm going to Washington University," Jim said bluntly.

Jerry Jr. doubled over in laughter.

Washington University in St. Louis, known as the Harvard of the Midwest, was not in the realm of possibility for any of the Kennedys. It was an elite school for brilliant Asian students, Jewish kids from the East Coast, and well-connected families in the Midwest. Mr. Kennedy made it clear he thought Jim had no place in such an institution. Jim saw an opportunity.

"I make you a bet," he told his father, "If I can get into Wash U, you'll pay for it."

"Yeah, sure. If you can get in, I'll pay for it," his father said, snickering.

Jim wanted witnesses, so he went and rounded up his aunts and brought them into the room, hoping that his father wouldn't back out if his wife's sisters, "who all hated my father with a passion," witnessed his bet.

"Tell them what you just told me," Jim admonished his father.

"Ha! I said I would pay for him to go to college if he got accepted into Wash U."

None of his aunts thought Jim could do it, but they would love nothing more than to see their brother-in-law lose a bet.

Six months later, in the fall of 1987, Jim Kennedy, a D-minus high school student who'd barely graduated, was on his way to the Harvard of the Midwest. "I couldn't believe it. I took the acceptance letter to my father like, 'Here's a great big fuck-you letter.'" Mr. Kennedy was incensed. "He didn't want me to succeed. He had zero invested in my success, and he was actually emotionally invested in my failure." Still, Jim's father did not back out of his promise.

Mr. Kennedy would now have to pay his son's very costly college tuition—about $20,000 a year at the time. As he was extremely frugal, it was now in his interest to hire his son as an employee of Kennedy Capital Management, even though the two were barely on speaking terms. That way, the tuition could be written off as a business expense. He told Jim that he would pay him a salary of $1,000 a month to do data entry. Jim would have to open the office at 5:00 A.M., work until 9:00 A.M., go to class, and then come back at the end of the day. Working twenty to thirty hours a week at Kennedy Capital, Jim enrolled for a full load of classes.

At Washington University, Jim declared his major to be business. He then decided to take all of his most difficult classes in his first year. His plan was to enjoy his senior year, focusing on the things he had learned to love: history, economics, the Greek tragedies, and geology. The cockiness that had accompanied him through the military and community college quickly vanished and was replaced by a near-constant fear. "I was trying, but I didn't know if I could succeed." His first semester at Washington University, he said, was like getting hit by a dump truck. The campus was full of prep school kids whose parents directed their lives for academic success.

"For the first time, I was afraid of failing. For all of my father's faults, he taught me that failing is okay. In fact, failing five times in a row is okay. You can you pick yourself up a sixth time and try again." But for Jim, this was a little different. "I did not see a second chance at something like this." Regardless, he was fully committed and would succeed or die trying. It was what he called his "Irish disease."

Jim had never taken algebra, geometry, or trigonometry, and the bare minimum requirement for Wash U business school was calculus. So he jumped right into calculus, barely keeping his head above water. His first-semester grades were Bs and a few Cs. But even as the semesters went by and his grades improved to As and Bs, he was still consumed by fear. He just couldn't shake the feeling that he was not cut out for Wash U. "I kept thinking, I don't belong here."

In the winter of 1988, in Jim's last semester of college, Jim's landlord decided the renovations were so well done that he threw Jim out so he could double the rent. Jim moved into another apartment, down the hallway from his old buddy Mark in a rough area of St. Louis. Mark was thrilled to have Jim nearby and very happy that he could still see glimmers of his old friend in the new, ultra-studious Jim Kennedy. One day, freaking out after hearing gunshots down the hall, Mark ran to see if Jim was okay. There he was at his desk, his handgun next to him. He'd been trying to study when two pigeons came to fornicate on the aluminum awning just outside his window, making the sound of fingernails on a chalkboard as they went at it for over an hour. So Jim shot them dead.

The two friends remained close, but it wouldn't be long before Jim's life would change completely, and Mark would have a second best friend to spend his days with.

It was the beginning of the school year in 1989. Eighteen-year-old Nina Abboud climbed the stairs of the massive lecture hall to her Western Civilization class. The stunningly beautiful freshman could barely concentrate on her classes her first day at Washington University. She had wanted to go away for college, but her Lebanese-born parents insisted that she stay in town and live at home. All her friends were off for their freshman year, while Nina was stuck in St. Louis. Even worse, she'd just broken up with her boyfriend. The thought of being a single eighteen-year-old college girl whose friends were scattered around the country seemed completely impossible. Nina not-so-subtly scanned the more than three hundred students in the auditorium, looking for prospects for a new social life. It looked dismal. Especially when "this guy walks in wearing a suit and holding a tape recorder. I'm going, 'Oh, jeez, look at that nerd. . . . This is going to be a long four years.'"

The nerd in the suit, of course, was Jim, who'd just arrived to the Western Civilization class a few minutes late, coming straight from Kennedy Capital. He was in his senior year but was taking freshman courses, since he'd finished his upper-division classes first. He looked up in the dozens of rows for an empty seat in the lecture hall. In the very top row, just about center, his eyes locked on Nina the exact moment she was checking him out. Nina quickly looked away, but not before she saw the first thing that actually intrigued her about the geek in the goofy suit; a very prominent scar that ran from his left temple down his cheek. Hmm, she thought. There might be something interesting about him after all.

Jim couldn't help but notice Nina; she completely stood out among the other coeds. Nina had wild waves of chestnut hair framing her face, and thick lashes lining her dark eyes. But Jim was not looking for distractions. The idea of a spending time with a girlfriend at this stage of his life had not even occurred to him.

The next day Nina wandered across campus, looking for her classes. When she found her Greek Tragedy class, she quickly sat at the back of the room. The class had already started when one more student rushed in late. "I thought, 'There's that nerd again,'" Nina says. "What were the odds we would be in two classes together?" And then later that same day, "It turns out we have study group for Western Civ together." By the third time she saw Jim, she began to see beyond the dorky exterior. Jim was in his last semester of business school, and Nina was a freshman in the School of Architecture. Typically they would never have crossed paths on campus, much less ended up with three classes together. At the time, Jim looked just like Michael J. Fox's character Alex P. Keaton from the television show *Family Ties*. And like Keaton, Jim was a staunch conservative.

After a few class days and a few awkward acknowledgements, Jim approached Nina in between classes. "He turned out to be quite funny and charming, in spite of his serious demeanor," says Nina. Although Jim found it impossible not to be attracted to Nina, he was by then twenty-five and felt much too old to be dating an eighteen-year-old. So he tried to set Nina up with his younger friend Don. Because they were all in Greek Tragedy together, the three of them began to hang out.

To Jim, Nina's carefree nature was totally alien. Her social life definitely came before her studies. She was very bright, and it seemed as if school was no effort at all. "So what if I have a paper due?" she would tell Jim. "I can do it in the morning."

"She just wanted to go out and have fun," Jim remembers. "I honestly didn't know how."

When Jim finally admitted to himself that he was falling for Nina, and Nina was obviously interested in him as well, he used his military adaptation skills and tried to have fun. "He totally fooled me," Nina jokes now. "Our first date, he's dancing with me in a cage in a nightclub on the East Side." Jim showed just enough of his spontaneous side to keep her intrigued. Unfortunately for Nina, Jim's wild nights did not last, and he quickly hunkered down exclusively on work and school. His cover as a typical Wash U student was soon blown as well. Nina only had a few dates with Jim before she got a taste of what her life would be like with one of what she would soon call "the other Kennedys." After Jim was late picking her up one night, Mark called. "Nina, Jim's in jail," he said. "We need to get $100 to bail him out."

Jim was on his way to pick up Nina in his new car when he was pulled over. Jim handed the cop his driver's license. The officer went to his car and then came back.

"Jim Kennedy, get out of the car. You're going to jail for armed robbery."

Jim put his arms behind his back and felt the unforgettable sensation of handcuffs on his wrists. "But I haven't done anything," he protested.

"That's not what it says here," the cop said, shoving Jim into the backseat of the cruiser.

Joe, Jim's second oldest brother, had used Jim's social security number and birth date the last time he was arrested. "So within forty minutes, I'm in an orange jumpsuit in a jail cell, and I'm so pissed. I can't get away from being a fucking piece-of-shit Kennedy."

Nina had no idea to how to react to the fact that her new boyfriend was in jail for armed robbery. She'd grown up in a wealthy suburb of St. Louis, the oldest of three. Her father, Nabil, a highly respected St. Louis doctor, and her mother, RoseMarie, a fashionable homemaker, nurse, and philanthropist, were strict parents who'd clashed with their daughter for many years. They wanted the best for their children, but struggled with reconciling their traditional Middle Eastern values with their "American" kids. Nina, her sister, Della, and her brother, Noel, were held to high expectations by their father. School was expected to be a priority. Fun, parties, and most definitely boyfriends were all things that should wait. How to get a boyfriend out of jail was something about which Nina had no clue. "I'm eighteen years old, living at home with my parents. I didn't have a hundred dollars," Nina says. Somehow she and Mark scraped together the money for Jim's bail and went to get him out. "I should have run at that point," Nina says. "but Jim had completely won me over. He was the most passionate, charismatic person I had ever been around."

When Nina introduced her new boyfriend to her parents, her father loved Jim immediately. "Jim came to pick me up for dinner. He met my father, and they started talking. I actually had to interrupt them after an hour so we could leave," says Nina. But no matter how much they liked Jim, it was still critically important to the Abbouds that Nina stay focused on school. They told her they didn't want her spending too much time with Jim. Nina responded by using the age-old tactic of American kids: sneaking out of the house and making up stories about her whereabouts. What Nina excelled at in flattery and charm, she completely lacked in lying and deception. "I was getting caught every other week," she says. After she was busted, scenes in the Abboud house would play out loudly and dramatically in English, French, and Arabic. Her father would fly into a fit of rage, and she'd scream and cry about the unfairness of it all, collapsing into a puddle of teenage tears. Nina's little brother was fifteen at the time, and the scrawny kid looked up to Jim and loved hanging out with him and his friends. In one of these explosive family scenes, Noel lobbied hard for his big sister. "You have to let her go!" Noel screamed. "I know she's going to marry him!"

As Jim and Nina grew closer, Jim's best friend Mark couldn't have been happier. Mark was Nina's "best girlfriend," and Mark loved Nina like a sister—the first time in his life he'd ever had a platonic girlfriend. When Jim was busy working or studying—a common occurrence—Nina and Mark would make plans and hang out. "Occasionally, we'd grab Jim and force him to take breaks for a little fun," says Nina. It wasn't easy to tear him away from his books. "When I met Jim, he was already in the mode of absorbing everything he possibly could about everything. Anything that interested him remotely, he wanted to learn about it—not like normal people who learn about something and then move on to something else." Nina believes Jim's drive was a function of how he was raised. "It's almost a survival thing: *I have to figure all this out by myself. No one is going to tell me anything about it.*"

When Jim graduated with honors from Washington University at the end of 1989, his first order of business was a return to Kirkwood High School. There was someone he needed to see. It was close to 5:00 P.M., and the place was mostly empty. Still in his suit and tie from work, Jim entered the main office and asked for Principal McCallie. "My secretary comes in and says, 'Mr. McCallie, there's a young man here to see you, and he says he's the worst student you ever had at Kirkwood High School,'" remembers McCallie, who'd been at the school more than a decade. "I said to my secretary, 'Let me give you five names of students who gave us real trouble.' One of the five names was Jim Kennedy." When Mr. McCallie walked out of his meeting, he was confronted with a mental disconnect. There was Jim, known to students, teachers, administrators, and principals as Freak, wearing a suit and tie. Jim remembers, "I said, 'Frank, it's Jim Kennedy. I just graduated from Washington University, and you are the first person I wanted to share that with.'"

Mr. McCallie quickly processed what Jim said. Suddenly he understood, and a well of emotion filled the room. "I thanked him for taking the risk of keeping me in school, and I told him I gave him credit for what I'd accomplished since then," Jim says. The two sat and talked for a short while. They shook hands and locked eyes, and in a brief moment their eyes said everything that remained unspoken. McCallie, with his lifelong passion for helping teenagers find their way, was incredibly touched that he'd been so remembered. "We had a wonderful talk. I was impressed that he came back to see me, and so pleased."

Jim continued working at Kennedy Capital. He'd passed a number of required exams and was working as a securities analyst, making a modest salary. Unfortunately, after graduation, he discovered that his education benefits from the firm—the way his father made good on his promise to pay for Washington U—were treated as income by the IRS for the tax years 1988 and 1989. His tax bill was more than his annual salary. It would take him years to pay it off.

In his free time, Jim read the *Economist*, the *New Republic*, and the *National Review* and biographies of current and past political leaders. His thoughts were evolving, and his hunger for knowledge became insatiable. When he found an ad for a master's program in political economics and public policy at Washington University in the back of *Economist*, he was sold.

While Jim's undergraduate experience had been laborious and stressful, graduate school—where he earned a scholarship that covered half the tuition—was pure joy. He studied under renowned economists, among them Douglas North, who won a Nobel Prize for economics, and Hyman P. Minsky, who, Jim heard, was rumored to be on the road to one. Minsky's ideas were consistent with Jim's evolving concerns about how massive and overblown the U.S. capital markets were becoming. He built his master's thesis around an idea he had to stimulate foreign investment into manufacturing and technology companies in the United States, and then decided that it was such a good idea that he needed to figure out a way to make it actually happen. In what would be his first foray into national politics, he got the ear of Illinois senator Paul Simon, who incorporated the language into a bill. The bill never made it to the Senate floor for a vote, but working with the U.S. Congress was exhilarating. The idea that he could change the course of his country for the better was the ultimate validation of Jim's new existence.

During Jim's two years of graduate school from the fall of 1991 to the spring of 1993, he deeply analyzed the American economic landscape at the end of the twentieth century. He wrote a paper on what he saw as a dangerous and growing divergence of interest between corporate management, shareholders, and boards of directors. A new philosophy filled corporate boardrooms and permeated the media so thoroughly that most people believed it was an actual law. A corporation is only responsible to its shareholders, it held.

"Then the stock market morphed into a hybrid casino and circus," Jim says. "It was an orgy of leveraged buyouts, leaving the surviving corporation saddled with debt and forced to cut costs, eliminate research and development, make massive layoffs, and divest assets. The only measurable purpose was to enrich the board, management, and shareholders at the expense of the corporation's future viability." His professor was so impressed that he openly questioned Jim's exclusive authorship and jokingly suggested that such ideas were incongruent with an academic career. For his final paper, Jim tried to evaluate the myriad of growing conflicts of corporate governance under this new cannibalistic form of capitalism. "It was nothing more than looting," says Jim.

And there was no one to stop the looters. The Grand Old Party that he'd once felt so akin to no longer seemed rational or sane or fiscally conservative at all. "The U.S. Congress was increasingly controlled by the capital liquidation strategies of Milton Friedman and the Chicago school of economics," Jim says. Friedman's economic principles morphed into a financial theology calling itself "free-market capitalism." Government needed to back off on regulations and clear the way for business to thrive. Lawmakers bent over backward to express how "pro-business" they were. Jim watched, incredulous, as this pernicious ideology found its way into every nook and cranny of academia, most think tanks, and both political parties.

In 1995 Georgia congressman Newt Gingrich, a former history and geography professor, became Speaker of the House. The Gingrich Congress championed Friedman's ideology, calling it a "Contract with America." Deregulating Wall Street, slashing government spending on research and social programs, and prodding Christian fundamentalists to infuse the party with antiscience fervor were the first orders of business. To Jim, Friedman's theories were not economics at all but what he called "an insane religion based on pseudo-science and faulty equations that lends all sovereignty to money." By Friedman's own definition, Jim says, "This ideology would allow for drive-through pedophile services and organ sales—because there would be a market for it." Jim was furious as he watched his political party being hijacked by forces of greed so strong and amoral that they were willing to sell out his country wholesale. "Adam Smith defined classical free-market theory, and it worked for two hundred years," he says. "In 1970 Milton Friedman turned the entire definition on its head." *Let the markets run free*. The Kool-Aid that Washington was lapping up and serving to the country would eventually create a catastrophic collapse that only a few, including Jim Kennedy, could see coming.

9

Indigenous Innovation

Across the hemisphere, in the People's Republic of China, Communist Party leaders had a diametrically different strategy for the future than Republican lawmakers in America. Science and technology were of paramount importance, at the top of the list for government funding. The future of China would depend on them, its leaders were sure. Chairman Jiang Zemin and President Deng Xiaoping already had many new policies in play to make progress in these fields happen, but one epic event sent China's technology revolution into overdrive: the U.S. annihilation of Iraqi forces in January 1991, a forty-five-day expo of technological supremacy called Desert Storm.

Long before that pivotal winter, Deng, who came to power in 1978 shortly after Mao's death, was a champion of modernization. Deng believed that robust government spending on science and technology would be essential to bring his country out of the disastrous hole dug by the Cultural Revolution.[1] (Between 1966 and 1976, Chairman Mao's policies banned his capitalist and bourgeois enemies from infiltrating the Communist Party, gutted China's research institutions, and eviscerated the country's economy.) Deng also believed that China's new communism should not mean zero involvement in global economics. For China to compete on the world stage, it needed enormous momentum, including great minds and the means to turn innovations into reality.

To implement his plan, Deng relied on a blueprint designed by four renowned Chinese scientists in 1986. All four held high appointments within the Chinese National Academy of Science and were members in good standing of the Communist Party. Not only were their collective scientific pedigrees exceptional but they also realized that science, innovation, and a nation's economy were inextricably linked. Seventy-nine-year-old Wang Ganchang was a major contributor to China's atomic bomb program in 1964 and chairman of the China Nuclear Society.[2] Chen Fangyun, seventy, a brilliant physicist, developed networks to control Chinese satellites. Yang Jiachi, a sixty-six-year-old specialist in space technology with a Ph.D. from Harvard, was the chief engineer of the Chinese Ministry of Aerospace Industry. Wang Daheng, at sixty-one, invented the first electron microscope and laser in China, and was known as the father of optical engineering. After studying and working in England for ten years, he returned to China to become the director of China's Center for Space Science and Technology. Wang Daheng credited his work ethic to his years in England. "I also realized how to set priorities, gain insights, and discover and analyze problems without the burden of old rules and constraints in solving problems," he said in a 2007 interview. "This is very necessary to scientific progress and development."

The four came together to pen a letter to Chairman Deng in March of 1986. Without a wide-ranging government program putting serious money into scientific research, they argued, China would continue to lag far behind Japan, Russia, and the West. The United States was solid on its Strategic Defense Initiative. Europe had a high-technology program called Eureka, and Japan was on a roll with its own tech efforts. They recommended a national program to monitor developments in all these countries, and suggested that China not only follow the progress of the exploding technological world but lead the way.

Deng was instantly receptive. The vision of these giants of science resonated with his reformist and capitalist ideas. "Action must be taken on this now; it cannot be put off!" he wrote on his copy of the letter. The following year, Deng announced what he called China's Program for Science and Technology Modernization, later dubbed Program 863 for the year and month—March 1986—it began.[3]

The plan was not secret. It was announced widely across media outlets. Seven sectors crucial to China's national security and economic competitiveness would get massive government resources: automation, biotechnology, energy, information technology, lasers, new materials, and space technology.

Some questioned Deng's plan as overly ambitious. He answered his detractors in a 1988 talk to physicists at Beijing's electron-positron collider: "China cannot afford to fall behind. China cannot afford *not* to be engaged in spite of the fact that we are poor. Because if you aren't engaged, if you don't develop in these areas, the gap will only become greater and it will become extremely difficult to catch up."

Deng was about to go full throttle on what would come to be called "indigenous innovation" in his awakening country. As China began to act like the America of the mid-twentieth century, putting money into research and championing a long-term plan for the future, Washington paid little attention. Intoxicated by exponential growth in the technology sector and an economy exploding forward out of a recession, many U.S. lawmakers championed American exceptionalism and looked at China as a poor backwater where 1.3 billion desperate Communists were willing to work for next to nothing for American companies. China would never rise from the scourge of Mao Zedong. It would be a tool for America forever.

While most people in America paid little attention to China's big plans, in a Philadelphia university, Program 863 must have been welcome news to a thirty-seven-year-old Ph.D. student named Jiang Mianheng. The freedom and funding to create science had not always been a given to the smart and dedicated engineering student, whose academic life had been preempted by Mao's Cultural Revolution.[4] As the son of Jiang Zemin, a prominent politician exiled by Mao for his reformist ideas, Jiang Mianheng was sent to the countryside outside of Shanghai in 1966 as part of an incipient national program called "rustication." He was only fourteen years old, just out of middle school. By 1968 the policy of rustication was countrywide. "It is necessary for educated young people to go to the countryside to be reeducated by the poor and lower-middle peasants," announced Mao, forcing an estimated 17 million other "educated youths" to abandon their studies in exchange for hard labor far from home.[5] The war on education raged not only against young people, the group that would become known as the "sent-down generation." It also extended to university faculties, which were purged of trained scientists and academics in favor of farmers and local revolutionary committees, breaking up the entire higher education system of China.

After Chairman Mao's death, Deng Xiaoping (who had been exiled by Mao to hard labor because of his reformist and "capitalist" ways[6]) took over the leadership of China and began dismantling the paralyzing constraints of the Cultural Revolution. It was a relief for Jiang Mianheng, who was finally able to return home and begin his university studies at twenty-eight years old. With a bachelor's degree in radiochemistry (the chemistry of radioactive materials) from Fudan University and a master's degree from the Institute of Semiconductors, Jiang arrived on the East Coast of the United States in 1986 to start his Ph.D. program at Drexel University in Philadelphia at the fairly advanced student age of thirty-seven.[7] The school had a long tradition of welcoming graduate students from China. One of the professors was a personal friend of Jiang's father.

At first Jiang stayed completely under the radar at Drexel. Like most other foreign graduate students, he lived on a stipend of $800 per month, received in exchange for a teaching assistantship.[8] With his wife and baby son, Jiang lived frugally in a three-story redbrick row house in West Philadelphia. He took classes and conducted experiments in the Department of Engineering. Professors called him hardworking, humble, and a brilliant researcher.[9]

One day in early June 1989, Jiang's quiet world exploded when Deng Xiaoping's military forces violently cracked down on antigovernment protesters in Beijing's Tiananmen Square. Hundreds of Drexel students protested in sympathy with students in Beijing.[10] Still trying to keep a low profile, Jiang was harassed and threatened for not supporting the uprising. It was common knowledge among the other students that Jiang's father, Jiang Zemin, the mayor of Shanghai since 1985, was a high-ranking politician in the Communist Party of China and supported the government assault on the protesters.

Soon after, at the end of June, the head of Drexel's engineering department, Bruce Eisenstein, called Jiang to his office to give him news that he was not at all expecting. With the world still in an uproar after Tiananmen Square, Jiang's father had been elected by party bosses to replace Deng Xiaoping. His father would be the head of the Communist Party of China. The news may have unsettled Jiang, but he remained focused on his work, and his professors say he refused police protection from Drexel and the

Philadelphia police. He spoke to a few reporters when the press called, but then stopped doing interviews altogether. He kept his head down and finished his dissertation.

The following year, Jiang Mianheng was working for Hewlett-Packard in California when Saddam Hussein invaded Kuwait. The United States was on the brink of its first major war since Vietnam, rattling sabers and calling for UN backing to help stop Saddam. Back home in China, Jiang's father, Secretary Jiang Zemin, was on high alert, waiting to see what the Americans would do next.

Also on the edge of his seat was U.S. Army major John Adams, a thirty-seven-year-old English literature professor at the U.S. Military Academy in West Point, New York. The six-foot-one army major was a commanding presence as he stood at the head of his class of about twenty young cadets in identical gray and blue uniforms on his last day of the fall semester in December 1990. His lecture room—housed inside one of the impressive Gothic Revival buildings on the fortresslike campus—was small and austere. The students were attentive as their professor wrapped up his lecture with his typical calm delivery. This semester, he'd enlightened West Point freshmen with Shakespeare, classic poetry, and twentieth-century American literature. It was his fifth semester at the academy, and after teaching at the University of Massachusetts in grad school, he was seriously impressed by his students' dedication. "West Point is a very different college environment," said Adams, an Army Ranger who also taught Ranger skills to the cadets during field training. "It's not like any other university. It's very intense, and cadets are focused on learning how to be military leaders from the time they get there."

That winter, the United States was in the midst of its largest military buildup since World War II, and being a cadet on the campus of West Point felt decidedly different than it had over the previous decades of the Cold War. Students on campus were abuzz with talk of the Middle East. It wasn't just among those in the military; the whole nation watched the massive U.S. deployment in response to Iraqi president Saddam Hussein's invasion of Kuwait the previous August. In Washington, President George H. W. Bush and his cabinet postured and strategized. A new twenty-four-hour news cycle brought by-the-minute images to nearly every living room in America. Even Thanksgiving dinner with Major Adams's parents in North Carolina turned to the topic of a possible Middle East war. "My dad asked, 'What do you think the chances are of you being deployed?'" Adams remembers. "I told him I thought the last place they would find soldiers to deploy would be from the English department at a U.S. military academy." Besides, he didn't feel like they would need someone with his skill set: after graduating from ROTC in high school in 1976, he trained as an aviator and flew dozens of surveillance missions during the Cold War between Eastern Bloc states and Western Europe. But he'd never been in combat. So Adams was completely shocked when three weeks later he got a call from Washington asking if he would consider deploying to Saudi Arabia. "I didn't count on the fact that I had been an aerial surveillance pilot," he says. "One of the needs they had was people who knew what targets looked like from ten to fifteen thousand feet." It struck Adams as odd that the military would ask him *whether* he would be willing to deploy. "When your hand is up at the beginning, and you swear an oath to the Constitution, you're saying you're willing to be deployed. It's not like they really have to ask me." But since they did, Adams went to the person with the most at stake in his absence—his wife. It was not an easy decision. Adams hated the idea of leaving his family. "I think anybody with a family realizes how precious the time is. When you leave your family for even short periods of time, you miss events in their lives that you really can't get back." He also asked his boss, the head of the English department. "My boss asked, 'What are your responsibilities?' and I said, "I think my responsibilities are to honor my commitment to the country. If the country needs my services, then I've got to go.'" Adams's boss was also supportive. "He said, 'We will obviously miss your services on the faculty, but you have a responsibility to answer this call.'"

In a frenzy to get his affairs in order, Adams had little time to reflect on his emotions. In truth, his heart was heavy. His seventy-one-year-old mother was beside herself. "She was really worried," says Adams. "She understood why I was going, but she was really, really against the war." In a desperate attempt to stop what she believed was a tragic mistake for America, Adams's mother wrote letters to every member of Congress, pleading they do everything they could to prevent a war. So while Adams was getting his shots and reading briefs and packing his bags, "She's just kicking out these letters because she's so overwhelmed," he remembers. "It was partly about me, but I believe she was taking a principled stand."

Within a week, Adams arrived at coalition forces headquarters in Saudi Arabia and met his staff of twelve. His first assignment was to identify Iraqi units to attack along the border of Iraq and Saudi Arabia. The data they used came from photographs taken by a U-2 spy plane flying at 70,000 feet. "In the images, we could tell where tanks were parked and the artillery pieces were. We were trying to kill tanks—specifically selected tank formations—for the areas where we needed to defeat their forces." When his team identified what they believed were the best targets for a first attack, Adams and the targeting staff passed their intelligence on to the air force for attack as Operation Desert Shield became Operation Desert Storm on January 15, 1991.

Jiang Zemin—father of Drexel student Jiang Mianheng—had only been the general secretary of the Chinese Communist Party for a year when the first bombs dropped on Saddam's forces. Jiang, who was also the chairman of the Central Military Commission, publicly condemned Saddam's invasion of Kuwait the previous summer and called for diplomatic measures. UN Security Council members passed one resolution after another against Iraq, but Saddam either wasn't listening or didn't care, and he gave no sign of backing down. Fed up after four months of an annoying juggernaut, the United States brought a resolution to the council that would allow "all necessary means" against Iraq if Saddam's forces did not withdraw from Kuwait by January 15, 1991. As permanent members of the UN Security Council, Chinese leaders could have vetoed the resolution. But, still reeling under sanctions from the West following the Tiananmen Square tragedy, they abstained. The resolution passed, and coalition forces prepared for war.

Behind the scenes, Jiang Zemin, eighty-seven-year-old Deng Xiaoping (who stepped down from Chinese Communist Party leadership in 1989 but still held the title of president of the People's Republic), and members of the secretariat were pulling for Saddam. State-owned media touted the supremacy of the Iraqi ground forces. Saddam's army, they said, armed with Russian, French, Brazilian, and Chinese-made military equipment, would crush the Americans on the ground, or at least hold them off until U.S. troops retreated with heavy casualties. The Americans were kidding themselves. It would be another Vietnam.

As Saddam continued to thumb his nose at the West, Operation Desert Shield moved over 3,000 fighter jets, bombers, and aircraft carriers into the Persian Gulf, Saudi Arabia, Oman, and the United Arab Emirates. Within five weeks, U.S. aircraft outnumbered Iraqi aircraft, and transport planes brought in 91,000 coalition troops. Even in the face of the largest offensive since World War II, China's *People's Liberation Army Daily* continued to argue that Saddam would deliver a successful counteroffensive. The American people didn't have "morality" on their side, they said. Saddam and Iraq, clearly the underdogs, were fighting a "people's war," along the lines of Mao Zedong's war dogma that still gripped China under its reformist leaders. The Americans would not prevail.

They could not have been more wrong.

On January 17, 1991, Chinese leaders watched, stupefied, as U.S. military forces launched Operation Desert Storm. The "shock and awe" played out like a sci-fi movie to a mostly delighted Western audience. In glowing green videos, alien warriors who could actually see in the dark pressed buttons on gleaming cockpit computers to secure their targets. In the first twenty-four hours, three hundred aircraft missions bombed military targets, obliterating the Iraqi air force and enveloping Saddam's army in an Armageddon of technological genius. Jiang and Deng, blown away by the supremacy of American airpower and convinced that the United States intended to "dominate the world," huddled with top military leaders and high-ranking Party officials.

Major John Adams was impressed with the coalition forces' ease in systematically killing targets. "We were fighting a war against a rather stationary defensive force," he says. "We could see them better than they could us. We had better tanks, better equipment." The U.S. tanks had optical devices that could see longer distances and guns that could shoot farther. "We could attack Iraqi military units with precision-guided munitions that really only came into widespread deployment in the eighties." The Iraqis, with no such capabilities, were completely unprepared for the onslaught. "We could hit them from fifteen thousand feet both with ballistic weapons and with missiles," Adams says. "So they wouldn't even know the missile was going to kill their tank or kill their artillery piece."

All of the new technology was amazing to Adams. "It told me that this is a new world. We had smart bombs and smart munitions. We had advanced targeting devices and advanced intelligence systems that were able to deploy both from space and from the atmosphere. We were using everything we could and all our technology to inflict maximum damage on Iraqi forces. And we were very successful at doing that." To Adams, it was clear that Operation Desert Storm was no Vietnam.

Days into the U.S. attack, the *People's Liberation Army Daily* (along with every other news outlet in the world, courtesy of the U.S. military publicity machine) showcased the sophisticated Western weapons systems: night vision equipment, aircraft with extended ranges and massive, guided payloads, and state-of-the-art reconnaissance and surveillance systems. Within weeks Saddam Hussein's troops completely lost the power to fight back. Men and machines were blown to smithereens, charred and mangled in a smoking graveyard desert.

"Over the course of about four weeks, we were successful in destroying almost all the Iraqi armor in the area," says Adams. "We were really effective on the ground and in the air. We saw massive casualties among the Iraq forces." American ground troops went in on February 22, 1991, for the final assault. "There was no contest," says Adams. "By the time we went in, it was all but over."

Chairman Jiang, known more for his knowledge of art, music, and economics than military strategies, had to come up with a plan, and fast. China's defense equipment was twenty years behind that used by coalition forces in the Gulf. It was time for a serious evaluation of his military.[11] Jiang and Deng called meetings and formed commissions. The two reformist politicians knew that China needed to completely overhaul its defense sectors and get serious about developing new science and technology.[12] Just one month after the end of the forty-five-day American offensive that annihilated the Iraqi army, the *Liberation Army Daily* reported Jiang Zemin as saying, "We must fight modern wars with modernized military. We must respect science and take serious views on our arms.

We should endeavor to enhance the level of our military's weapons and equipment as much as our national budget allows."[13]

Jiang Zemin wasted no time. The state council immediately ordered twenty-seven high-tech development zones in cities across China.[14] Several billion dollars' worth of computer equipment filled state factories. Jiang knew that China would remain a backward country if its population did not innovate on its own. He called for a policy of "indigenous innovation," and vastly increased spending on science and technology. But both Deng and Jiang knew that even a massive spending plan would not be enough to dig China out of the hole left by the Cultural Revolution. Jiang was a trained engineer. In fact, eleven out of twelve members of the secretariat were engineers. (Western political scientists would soon tag a member of the ruling class with a technical background a "technocrat.") The Chinese technocrats knew that developing technology from scratch took years, sometimes decades—and a lot of money. They had money. They didn't have time. While many new Chinese discoveries had been made since the start of Program 863, they were certainly not at the rate or quality needed to compete on this new dangerous world stage.[15] They needed a plan B.

Jiang Zemin and his fellow leaders were not only well versed in scientific topics but their shrewdness also made them keen observers and investigators of all facets of global societies. They studied the progress of other nations' economies and scientific advances. They analyzed the ideology of other nations' governments and people. And in so doing, they developed a concerted plan—a detailed and comprehensive course of action to get precisely what China needed to become a dominant player in this new world order. One item on their agenda (which was, of course, not publicly stated) was espionage.[16] An intelligence initiative continued, and as they had for decades, China inserted operatives into the world's secret weapons labs. This method was always dangerous, fraught with problems that could once again isolate China economically from the West. Jiang Zemin knew that there was a much easier way. Wouldn't it be better to just buy the technology from those who created it?

At the time, Jiang Zemin probably felt the acquisition part of his plan was a long shot. For the United States and its Western allies, an international agreement had been in place since 1948 that kept defense technology—and other technologies with both defense and commercial applications, called dual-use technologies—out of the hands of enemy states. The Coordinating Committee for Multilateral Export Controls (COCOM) had been a big success during the Cold War.[17] Russia and its allies fell far behind the West as member states worked together to keep sensitive equipment away from the rising Communist regimes. If equipment was stolen, spare parts and service were impossible to get, making it too chancy to build manufacturing plants around smuggled technology.

For Peter Leitner, a senior strategic trade adviser for the Department of Defense, COCOM was a wonderful tool that kept America and its allies safe, and he was its proud gatekeeper. Since 1986, his charge had been to ensure that companies couldn't sell their dual-use technologies to enemy foreign countries. He frequently flew between Washington, D.C., and Paris, where COCOM countries met to discuss how different export situations should be handled and particular items should be restricted. And because of the tight rules, China was denied virtually all access to high-performance computers that could build missiles and advanced nuclear bombs.

"During the Reagan administration, we were the guardians of American technology as it related to military capabilities," says Leitner. At that time, he and other DoD advisers operated under the policy that "technology is born embargoed," he says. "The assumption was that everything is controlled unless it's specifically exempted from control." What turned out to be the biggest hurdle and frustration for Leitner was actually his sister agency, the U.S. Department of Commerce. "Commerce is in favor of giving virtually anything to anybody at any time," says Leitner. But though Commerce seemed to rubber-stamp everything, under Reagan and George H. W. Bush the DoD had a heavy hand, preventing a lot of technology from getting into enemy hands.

That was about to change.

It must surely have been music to Jiang Zemin's ears when, almost immediately after coming to office in 1992, President Bill Clinton argued that with the Cold War over, military and political threats around the world no longer existed. *Let's get down to the business of business.* With the help of newly appointed commerce secretary Ron Brown, Clinton's first order of business was to get rid of the pesky COCOM, which, the White House argued, was bad for American corporations. The administration pushed member countries to dissolve the committee, and with the major superpower ready to back out, they agreed. The administration promised to work on some new guidelines and develop policies to take COCOM's place, but there was no serious movement in that direction. That same year, another stopgap to losing sensitive technology called the Export Administration Act expired, and Congress failed to agree on how to keep it afloat. Now there was very little way to penalize companies that sold critical technology to any countries they chose, even dangerous enemies of the United States.

Suddenly, Peter Leitner's world was upside down. His new bosses, Secretary of Defense William Perry and Undersecretary of Defense Ashton Carter, had been appointed by Clinton and had the president's take on global trade. "They were totally in favor of selling virtually anything to the Chinese," Leitner said. Defense Secretary Perry called controlling dual-use technology a "hopeless task" that "only interferes with a company's ability to succeed internationally if we try to impose all sorts of controls in that area." Assistant Secretary Carter proposed giving Pakistan the special electronic locks used in nuclear warheads and supplying nuclear fuel to India. "So our policy changed in the DoD, basically weakening our opposition to Chinese acquiring critical U.S. high technology," Leitner says. "Officials in the DoD were trying to decontrol as many things as possible. So the safety net, the export control regime, everything that was there to protect U.S. industry and U.S. technological advantage, was being dismantled." Objecting to the recklessness of the new policies, Leitner swore to continue doing the job he believed he was hired to do. He was sure that selling dual-use technology would end in disaster. Perhaps Bill Clinton and his new DoD bosses didn't get it, but Peter Leitner knew that virtually every bit of weapons and defense equipment incorporates dual-use technology.

So did Jiang Zemin.

When Jiang Zemin saw the big For Sale sign go up for dual-use technology from America's most innovative technology and defense contractors, it must have warmed the cockles of his heart. China's plan for a technologic takeover was bold and direct: invite the major military and tech contractors to open their manufacturing plants in China. Give them free rein with cheap labor. Hold them to no environmental or human rights standards. Give them half the tax rate of Chinese corporations, and dangle before them 1.2 billion consumers and a government with money to purchase their products. What would the corporations have to do in return? Simple. Share their technology. Open up their files. Teach Chinese scientists their methods via partner companies. To Jiang Zemin, it was a no-brainer. If the American, European, and Japanese companies went for it, with zero capital invested, China would be handed the most important secrets and inventions of the twentieth century.

Jiang Zemin's strategy was an instant success. One U.S. major multinational corporation after another lined up to move their manufacturing plants to the People's Republic of China. Motorola, 3M, and Philips were some of the first to go. Japanese and German companies jumped on the bandwagon, as Volkswagen and NEC (a Tokyo-based IT company) began production on the mainland. None of the relocating corporations received any guarantee that their intellectual property would be protected under U.S. or international patent laws.

One of the most enthusiastic beneficiaries in China's global tech buyout was none other than Jiang Zemin's son Jiang Mianheng, who returned home to Shanghai in 1992 with a deep knowledge of Western scientific education and technology companies. For smaller tech manufacturing corporations, moving to China did not offer great benefits. But many of those companies had things that China wanted. And in America, it was open season on the best technology that money could buy. Chinese state-owned companies bought U.S. companies (some of which held the patents for discoveries funded by the U.S. government) and then systematically packed up whole operations and deposited them in Beijing, Shanghai, and Shenzhen. In 1995 one move in particular created a vacuum that would gut the future of technology manufacturing and innovation in America. The company was based in Anderson, Indiana, and named after its one and only patented product, a product integral to almost every single computer being made on the planet. That company was Magnequench.

10

Kennedy Capital

For Jim Kennedy, the 1991 attack on Saddam's forces offered a dark window into the American political machine he'd aligned himself with. He kept abreast of the news by reading the *Wall Street Journal*, the *National Review*, and the *New Republic*. "I knew that 'premature babies tossed from their incubators by Iraqi soldiers' and 'imperial rape rooms for Saddam and his sons' were total fabrications to get the U.S. public behind the action," he says. "I found it morally repugnant, but I understood the political considerations behind it," he says. Although Jim was an avid supporter of George Bush Sr. and impressed by his diplomacy in getting so many Arab countries to support the attack, he believed that the invasion was an obvious power grab for control over oil, and nothing more. For the first time since he'd left the army, Jim was glad that he was not a member of the U.S. Special Forces. He wanted no part of Desert Storm. He was very happy to be home in St. Louis, engaged to the beautiful Nina Abboud, and living a life he could never have imagined.

Jim's experience at Washington University proved that he could be competitive with the world's academic elite, and it gave him a new, unbridled confidence. Somehow he was still able to work alongside his father, and he became a securities analyst at Kennedy Capital. His job was to research companies and choose which investments would make his clients the best returns. In the early 1990s Wall Street followed the conservative business model that Jim had studied in school. As businesses grew and produced earnings, they invested in their own companies. Investors understood that if a company spent a lot of money on a large-scale project, it would increase the company's future earnings. Kennedy Capital made its clients money by betting on companies that looked like they would grow in the future. "At the same time, we protected our clients by seeking 'value' in the companies we invested in, ones that had low price to earnings ratios [the stock's price divided by its earnings per share over the past year] and high book value [the total net worth of a company's assets]," says Jim. "My father described it as the John Templeton model of buying 'garage sale' stocks cheap."

They may have been conservative money managers, but Richard Todaro—who came to work at Kennedy Capital as a twenty-year-old intern—remembers that both Jim and Mr. Kennedy (no one ever called Gerald Kennedy by his first name) were happiest when they got wind of a wild investment opportunity. "A firm would come in—it could be a gold company who said they found gold. Maybe the average person would say, 'Hey, no. I'm not interested. Just leave me alone and shut the door.' But Jim was a big-picture guy. He and his father would hear the story. They wouldn't necessarily invest or buy into it, but they were willing to hear all the stories."

As a junior analyst, Todaro was always impressed by Jim's way with clients. "We would meet with CEOs and CFOs all day long. I would have my suit on, and everybody would be all stuffy. And Jim would come in and he would make jokes with the people. If we went to dinner, he would try not to bring up anything snooty. He was always down-to-earth about everything." It was what Todaro loved best about his mentor. "I could see that it set the tone for the room. You could see people start to relax," he says. "You have to have a lot of confidence in yourself to do that."

Todaro never saw Jim and his father in conflict at work, and remembers Mr. Kennedy as a decent and fair boss to most of his employees. Unfortunately, the relationship between the father and son did not improve. "Mr. Kennedy would actively do things to sabotage the projects or the commission that Jim was working on," says Nina, who also worked at Kennedy Capital as a bookkeeper at the time. "Jim started marketing the firm in the early 1990s. He helped build that business. Then every time that Jim would be successful, which was a success for the company, all of the sudden there would be a renegotiation; 'I'm not going to pay you that . . . I didn't think you were really going to get the client.'" Jim says that because his father never thought he would succeed, "He would offer me the same promise that he would give to someone else. Then, when I was successful, he mentally could not deal with treating me the same as an outsider."

In the early 1990s, Jim started making a lot of money for Kennedy Capital. He landed its first institutional client, the Missouri State pension account. He followed with the Rabbinical Society out of New York. Then he landed the state of Michigan. Between those three clients, he brought in $300 million under Kennedy Capital's investment. "I should have been paid about $1 million per year in commission," he says. "That is what we paid outside marketing agents for the same results. I got paid about twenty-five percent of that—which was still great. My father screwed me, but it didn't bother me. I never wanted anything from him." Plus, he says, "I never cared about money, to a fault."

By the mid-1990s, the inner workings of Wall Street had quietly changed. Historically, investment banks operated on the bank partners' money. "Every bet, every investment, every initial public offering, could enrich the owners or break them," Jim says. "So the partners avoided anything that could go bad or roll back on them." When the investment banks "went public," everything changed: Merrill Lynch went public in 1971, Morgan Stanley in 1986, Lehman Brothers in 1994, and Goldman Sachs in 1999. With no personal capital at risk, bank owners (the partners became "owners" in the IPO process) could now bet on anything. The owners were protected by a corporate veil because a public company could not be sent to jail. Corporate stocks became vehicles for wild profits, not a reflection of a company's actual worth. Institutional investors—the mutual funds, insurance companies, and hedge funds of the world—pushed management and corporate boards to spike short-term profits. Investment banks brought ever-sketchier stock offerings to Wall Street. Accounting gimmicks and offshoring ran rampant. To Jim, it all violated the first rule of finance: accountability. "Nobody was responsible for what they sold anymore . . . or whether it was good or bad or real or fake." To justify huge fees paid to the new "professional managers" and "asset consultants" who ran the hedge funds and mutual funds, returns had to be in the double digits. "If you understand basic math," says Jim, "this is not economically sustainable for any business model over time."

"Wall Street established an expectation of insane growth, year after year, in perpetuity," says Jim. To meet that growth, companies would make promises that could never be fulfilled. "One bullshit company after another got launched based on double-digit earnings expectations," says Jim. "Then, somehow, a company grew sales exponentially with no corresponding earnings any time in the foreseeable future. For example, Amazon.com Inc. has a long history of accumulated losses, but a very high stock price. Amazon's stock value is based on its spectacular sales growth. But its growth is extremely high because Amazon sells products below cost, so it has no or low earnings." It drove Jim crazy.

"The markets found they could make more money dissecting things than building them," says Jim. Assets were dismantled and sold off. Manufacturing was offshored to China. Owners laid off workers, terminated R&D divisions, and raided pensions to pay onetime dividends to shareholders. Jim watched, sickened, as the cannibalism allowed capital markets to meet unrealistic return expectations by generating one bubble after the next. "If you just consider the history of the biotech bubble and the dot-com bubble, normal financial parameters like earnings or book value became irrelevant."

To Jim, Wall Street's new model was completely contrary to reality. The fallout was exactly as his favorite graduate school professor, Hyman Minsky, had predicted in the 1980s: When investment decisions are divorced from consequences, bad things follow. Jim was beside himself, his father even more so. Jim's clients expected the same returns they were seeing on Wall Street.

Kennedy Capital's model was to stick with companies that had tangible assets, low price-to-earnings ratios, and low debt. By the mid-1990s, "Everything that we were doing right as an investment firm made us look stupid in the short term. We continued to purchase value stocks even though the market had gone nuts." After a "correction"—when the value stocks recovered and the others languished—Kennedy Capital would be back on top. "But that was little consolation in the face of one bubble after the next with people making obscene money based on nothing real. The market rewarded people and institutions for abandoning all of the traditional metrics of investing," Jim says. It was all too much for Jim's father and his conservative sensibilities.

In the midst of the dot-com bubble, Mr. Kennedy came into the office one day. "He was ashen," remembers Nina. "He lost a tremendous amount of money because he shorted Amazon.com." The new Internet start-up was a ridiculous concept to the old-school investor. "Because the company grew sales by giving away product below its actual cost, he knew that Amazon should go to zero," says Jim. "He was a fundamental guy. He believed that a company had to make money to be successful. But in the dot-com bubble, the more ridiculous the company, the more its stock price would increase. For my dad, the whole world shifted at that point."

Kennedy Capital was at risk of falling behind leading indexes, which freaked out their clients, who were used to seeing stellar returns. Mr. Kennedy felt pressure to take risks that he never would have before. "Eventually, something was bound to go bad," says Jim. And it did. The firm bought into a small nutrition company posting fantastic profits and making impressive medical claims. Kennedy Capital locked in more than $2 million of its client's money before Jim learned that the company's CEO was hiding a felony conviction. He was legally unfit to run the company. When news got out, the stock price tanked. Jim gathered together the company's shareholders to oust the CEO and replace him with a qualified person. His plan backfired when the CEO turned the tables and sued Kennedy Capital. "I swear to God, I had a nervous breakdown when that happened," says Jim. "I lost my mind. I don't know how Nina put up with it." Jim remembers it as one of the worst years of his life. In his mid-thirties, Jim was so stressed out, he had one car accident after another. "He increasingly didn't like going to work," says Nina. "Things didn't make sense to him. He was miserable."

Exhausted and wanting out of the business, Mr. Kennedy was actively trying to sell Kennedy Capital. He had several offers, but they were always the same. "No one was actually offering to 'buy' the business in the traditional sense, they were only willing to pre-pay based on the future income of the business," Jim says. "At this point, I was sick of it, emotionally burnt, and I had no interest in leading the company." Jim began to structure a deal. It was the only way he could think of that would get his father and mother a well-earned payday and protect the team that had been with Kennedy Capital for more than a decade: an employee buyout. While Jim worked out the details, Mr. Kennedy and Jim's mother left for a cruise. A week later, on March 17, 1999, Gerald Kennedy had a massive heart attack aboard a ship in the middle of the Arabian Sea.

"He died on St. Patrick's Day, one week after his sixty-fifth birthday," says Jim. "I felt sorry because no one loved him. There wasn't a word of remorse from any of my siblings. He was a high-energy person. He had unlimited ambition. He just didn't know how to deal with raising children." Jim knew that his father had grown up in a similar environment, "saturated in hatred and jealousy." At the funeral, one of Mr. Kennedy's siblings insulted him at the podium about his "love of money."

Afterward, "Everyone looked to Jim and wondered if he wanted to take over the company," Nina remembers. While Jim was trying to stabilize Kennedy Capital, out of nowhere, the company's accountant made a run at a takeover. "It was a disaster. People threatened to walk," Jim says. Dealing with it all was making Jim physically ill. "I don't drink. I don't smoke," he told Mark Haselhorst. "And every day I'm driving home and my chest hurts.'"

Even more painful to Jim was Wall Street's new way of doing business. "I just hated it. I couldn't find anything real. If you actually understand how the stock market now works, you understand that it's a liquidation machine. It's all a system for divesting the future of a company for near-term profits, and it's really bad and very destructive for America." Jim spent nearly fifteen years in the financial industry. "During that time, I saw horrible things happen. I found it to be utterly depressing." By the end of the 1990s, Jim had no faith in the U.S. equity markets and growing concerns about the long-term viability of the country. He was sure that the "Minsky Moment" (the market crash his professor had taught him about) was right around the corner.

Nina, who had just given birth to their first daughter, Dravin, was very concerned. "Jim looked twenty years older than his thirty-seven years. He was gaunt and pale." The couple went to Nina's parents' house several nights a week for dinner. Her once-gregarious husband didn't socialize or interact with anybody at all. "I realized he was becoming very depressed. It was like pulling teeth to get him to do anything or go anywhere with me. He wasn't the guy that I knew."

Frightened and needing advice, Nina went to the one person who had always counseled her. "My dad is a very practical, conservative guy," she says. "He always thought about things in context of long-term consequences. He's a planner who believes that sometimes, present-day sacrifices are necessary to ensure a stable future." So Nina expected her father to give her the usual sensible advice—to tell her that with a brand-new baby, Jim needed to keep his job with Kennedy Capital. Instead, "he told me, 'You know what, Nina? Jim looks terrible. It's clear that he's miserable. He will die if he keeps it up. He needs to leave Kennedy Capital before the job kills him.'" Nina was shocked.

Soon after, with her mother's encouragement (and RoseMarie's desire to have her new granddaughter all to herself), Nina left for her annual solo five-hundred-mile bike ride. Hour after hour through small towns and across rural Missouri farmland, Nina brainstormed on how to help Jim out of his depression.

Jim was home in St. Louis, paralyzed and clueless at how to uncouple himself from Kennedy Capital, when Dr. Abboud came to see him. "You need to quit your job," Abboud told his son-in-law, "or you are going to die. I can see it. It's killing you." Nabil also gave Jim a reassurance factor that he'd never considered from his own family. He told Jim, "You've got us. We will not let you fail."

Jim thought about it through several sleepless nights. Nabil was the father that Jim never had, and he would never have considered quitting his job if he didn't have the support of his in-laws. The mere thought of leaving Kennedy Capital gave him such a feeling of relief, he knew immediately it was the right thing to do. He couldn't wait to call Nina and give her the good news. In a campground with spotty cell service, Nina got the call. "He tells me, 'I'm quitting. I'm not going to do it anymore. I'm done.' I thought, 'Oh, gosh . . . We can't do that! How are we going to take care of this baby?'" Her mind raced for a few minutes. "We'd talked about it a lot, and I'd been thinking about it. But the reality was scary, even though I knew there was really no other option. I thought maybe we'd figure it out, stay with Kennedy Capital and get through it." Ultimately, Nina knew leaving Kennedy Capital was the right thing to do for her husband. "I just tried to be supportive. As soon as I said, 'You know, honey, it's going to be fine, we will make this work. We have the support of my family. Nobody is going to tell you that you can't or that you shouldn't. In fact, we're all in this together.' As soon as that happened, you could almost see a weight being lifted."

Nina felt relieved too. The "nerd" she'd married was truly her best friend, and she was his biggest fan. "Jim always has an idea and a plan," she says. Jim would offer Nina many more sleepless nights and heart palpitations long into their future, but she would never, ever doubt that Jim would find a solution in the end. "We had a tiny house and not many expenses. I wasn't worried that we'd end up homeless. I knew that my family would support us if we fell on hard times. I just wanted to see Jim doing something productive and moving forward. I wanted to see the passionate Jim Kennedy pursuing something that excited him again."

In no time at all, Jim had a plan.

In the early 1990s, when he was first earning a lot of money and needed a tax shelter, Jim and Nina bought a piece of property eighty miles southeast of St. Louis in the St. Francois Mountains. For months he and Nina hiked around the three-hundred-acre parcel over and over until they found the perfect place on a hill, where they could see the sun rise through the oak trees on a ridge across the valley. "What are you going to do with all this?" Mark asked when Jim brought him up to visit. "I'm going to print business cards," he said, "that say: Jim Kennedy, Cowboy."

11

The Acquisition

Soon after the dissolution of COCOM in 1994, John Croat arrived for work dressed in his usual suit and tie. The creator of the Magnequench magnet was now a director for the company that bore his invention's name. Today, he would represent the research arm of the company to a potential buyer. "General Motors got into financial difficulty at this time," Croat says. "They were running out of cash, and they started selling off anything they considered non-core because they wanted to go back to the business of making cars." One business that they decided to sell was Magnequench. On a number of occasions over the past year, Croat and the other senior managers at Magnequench had met with prospective buyers. The meetings were set up by General Motors headquarters in Detroit. Croat and other senior managers gave detailed three-hour-long presentations on the entire business, including attributes of the melt-spun magnetic powder, existing and pending intellectual property, sales prospects, and current financials. Months later, Croat and his fellow executives would hear that there was an offer on the table. Several American companies made bids, but nothing stuck.

This day was different. The potential buyers were from China. Croat remembers the two, a man and a woman who listened intently and then left. Several months later, Croat learned that Magnequench had been sold. Managers and employees heard most of the details of the sale in the media, rather than from General Motors. On June 28, 2005, PR Newswire reported that "General Motors Corporation (NYSE: GM) has agreed to sell its Magnequench product line and manufacturing plant to Magnequench International, Inc. (MQI), a Delaware corporation owned by two Chinese state-owned companies and a U.S.-based investment firm. The sale is subject to approval by the U.S. government."[1]

The U.S. part of the trio of companies purchasing Magnequench with an offer of $70 million—$25 million more than any other offer, according to Croat—was led by Archibald Cox Jr., a Washington insider and the son of the famous Watergate prosecutor.[2] With Cox's Beltway connections, there was little problem getting approval from the government for Magnequench's sale to a majority Chinese company. A deal quietly passed after Cox promised the Committee on Foreign Investment in the United States that he would not move Magnequench's production equipment or jobs from the United States for ten years. To the unionized workers, Cox committed to investing money in the plants and keeping production lines going for at least as long. General Motors' executive vice president, J. T. Battenberg III, also tried to calm workers' fears in a media statement: "This transaction will maintain Magnequench's operations and jobs in the United States and promote future growth. Its association with San Huan, CNIEC [the two Chinese companies], and Sextant [Cox's company] will strengthen its ability to compete in global markets."[3]

Peter Leitner—who got wind of the sale in his office at the Pentagon—was alarmed that two Chinese companies could buy such an important technology company from the United States. He believed the deal would never have happened without the help of a Washington insider like Archibald Cox Jr. Most disturbingly, Magnequench was not just going to Chinese entrepreneurs in a new socialist-capitalist system. The U.S. company—its products so integral to weapons systems—was being purchased by two sons-in-law of Deng Xiaoping. The former Communist Party leader was now in his nineties, but he had known for a very long time how important rare earth magnets were for the future of every developed country in the world.

And it just so happened that the purchase of Magnequench fell perfectly into Deng's master plan. In the 1980s and '90s, working to bolster his country's disastrous economy, Deng commercialized its monolithic government-controlled industries, giving partial control to private enterprise. Many profitable enterprises went to family members of high-ranking party officials. Deng handed over control of China National Non-Ferrous Metals Co., Ltd. to his daughter, Deng Lin, and son-in-law, Wu Jianchang. In doing so, he gave them a monopoly on the sale and trade of all nonferrous metals—that is, all metals except for iron or iron alloys. It was a gift that came with enormous influence and earning potential.

Since Desert Storm, Deng and Jiang Zemin had been orchestrating China's technological revolution on every level from basic resources to manufacturing. Deng was delighted to learn of the massive neodymium and other light rare earth reserves available in China, as a side product of a giant Inner Mongolian iron mine called Bayan Obo. Well into his eighties, Deng was pushing China to adopt a global strategy for strategic resources, and spoke publicly about rare earths in 1992. "The Middle East has its oil. China has rare earth," China Public Radio reported Deng saying while commenting on the mineral riches of Inner Mongolia.[4] But China's technocrats knew that having the resources was only part of the equation. China needed to know what to do with them to really gain control of the market. That's why his sons-in-law made a beeline for Magnequench when it went up for sale three years later. With Archibald Cox greasing the wheels in Washington, they bought it, lock, stock, and barrel, and prepared to move it to China.

"The company was supposed to stay in Anderson, Indiana, and it was supposed to stay an American company," says John Croat. "I had a feeling that it never would. I suspected they wanted to move to China at the time they bought it." Almost every salaried employee, Croat says, was eventually fired or left in the following few years. The 225 hourly workers all lost their jobs.[5] Just eleven years after Magnequench broke ground, Anderson, Indiana's crown jewel of technology manufacturing was no more. The better mousetrap now belonged to China.

The next move for the new Chinese-owned Magnequench was to buy the only company in the United States that held a license from Sumitomo Corporation to produce Dr. Masato Sagawa's "sintered" neodymium-iron-boron magnets. Now Cox and his two Chinese partners had everything they needed to completely gut U.S. manufacturing of permanent rare earth magnet technology and set up shop in China.

Two years later, economic relations between China and the United States were decidedly warm and fuzzy when Jiang Zemin came to the United States for a 1997 visit. In good spirits, Jiang Zemin brushed off human rights protesters calling for a free Tibet and denouncing Tiananmen

Square. "China's market is open to you," he cheerfully told two hundred corporate CEOs at a dinner in New York City.[6] He toured the IBM headquarters in Manhattan. He visited AT&T and Lucent Technologies in New Jersey.[7] He rang the bell at the New York Stock Exchange. Days later, in Southern California, Jiang Zemin sat comfortably in a Buick Regal, the same model that would begin production in China the following year.[8] He stopped by Hughes Electronics Corporation to see the satellites soon to be delivered to the People's Republic of China.

By the late 1990s, a handful of people in Washington, D.C., began to feel that the lucrative back-scratching between the China and America's top technology and defense companies might be coming at a higher price than anyone recognized. Few in Washington wanted to discuss it. The stock market had rallied like never before, and American multinational corporations touted record profits. But to a few lawmakers and analysts in Washington, China seemed to be gobbling up the most prestigious and important American technology at an alarming rate. It was a particularly disturbing issue to California Republican congressman Christopher Cox (no relation to Archibald), who called for a bipartisan commission to investigate.

The *Report of the Select Committee on U.S. National Security and Military/Commercial Concerns with the People's Republic of China* was released publicly in May 1999, after 30 percent of it had been redacted by the Clinton White House.[9] What became known as the Cox report read like a thousand-page playbook that could just as well have been titled *How to Take Over the Military Technology of the World*. The revelation that China had been infiltrating poorly secured American weapons labs since the 1970s sent shock waves through Washington and the Western world. Oak Ridge, Los Alamos, Lawrence Livermore, and Sandia laboratories had all been targets. And in 1995 a Chinese national came forward claiming to have stolen the plans for the W88, the most advanced thermonuclear bomb ever developed. Worse yet, Clinton's national security adviser, Anthony Lake, who found out about the espionage in 1995 from the CIA, failed to tell Clinton. Because of what many officials described to the *New York Times* as "miscommunication, bureaucratic inertia and outright bungling by several agencies," it took until 1997 for Clinton's new national

Security Advisor, Sandy Berger, to give the president the bad news.[10] Along with these startling revelations, the Cox Report delivered dire warnings:

> The stolen U.S. nuclear secrets give the PRC [People's Republic of China] design information on thermonuclear weapons on a par with our own. Currently deployed PRC ICBMs [intercontinental ballistic missiles] targeted on U.S. cities are based on 1950s-era nuclear weapons designs. With the stolen U.S. technology, the PRC has leaped, in a handful of years, from 1950s-era strategic nuclear capabilities to the more modern thermonuclear weapons designs.
>
> These modern thermonuclear weapons took the United States decades of effort, hundreds of millions of dollars, and numerous nuclear tests to achieve.

While the shocking espionage revealed in the report received a fair share of press, what really should have gripped the White House and the Pentagon and reverberated through the American business and manufacturing communities was what China was doing through strictly legal maneuvers: "Since the early 1990s, the PRC has been increasingly focused on acquiring U.S. and foreign technology and equipment, including particularly dual-use technologies that can be integrated into the PRC's military and industrial bases . . . These pervasive efforts pose a particularly significant threat to U.S. exports control and counterintelligence efforts."

Peter Leitner hoped the explosive information in the report would reverse the tide and bring back a stronger export control system. Instead, he says, "Nothing happened, and there was virtually no follow-up." The report was never assigned to a permanent committee, nor was any action taken to tighten trade laws. Eight months later, a new Congress came to Washington along with a new president. What should have been a clear wake-up call to the government of the United States was summarily ignored and filed away in the United States Library of Congress.

With seemingly no one paying attention—even when he went to Capitol Hill to testify to Congress about the problem—Leitner continued his work at the Department of Defense. When he denied the transfer of sensitive technologies, his veto was almost always rejected by Commerce Department representatives, who had the final say. His last hope was that the second Bush administration would reverse the course before America was completely gutted of its technological base. George W. Bush campaigned on the idea of tightening up exports, but things got worse. In April 2002, pushed by the aerospace industry, Congress convinced the State Department to ease restrictions on satellite sales.[11] Yet in his 2003 State of the Union address, Bush once again promised to "strengthen global treaties banning the production and shipment of missile technologies."[12]

Two months later, and eight years after the sale of Magnequench, the paperwork requesting to move the company's patented melt-spinning technology to China landed on Leitner's desk in the Pentagon. "They were basically trying to replicate the production line in China, and they were coming in for export licenses because those devices were export controlled." Leitner believed that if China controlled access to the magnets and the rare earth materials they were composed of, it could hold U.S. industries hostage with blackmail and extortion. By now, Magnequench magnets were powering thousands of military applications, including smart bomb technology. "I was saying, 'There is no way we're allowing you to do this. You're basically shifting an entire U.S. sub-industry to the Chinese. This is a critical and strategic material. These rare earth magnets are used in our weapons complex. No, I'm not going to allow you to do it.'"

There were two federal lawmakers on Leitner's side. Both were from Indiana, where the last Magnequench plant in Valparaiso was packing up for its final move to China, three years earlier than the new Chinese owners had promised. With a few hundred constituents about to lose well-paid union jobs, Democrats Evan Bayh and Peter Visclosky tried to get the attention of President George W. Bush:

> March 6, 2003
> Dear Mr. President:
>
> We are writing to request your intervention to prevent a potentially serious loss of national security production and jobs in the United States.
>
> We have been informed that a facility owned by Magnequench Inc. in Valparaiso, Indiana may soon close and move its operations to China.

The Valparaiso facility produces rare-earth magnets used by the U.S. military in precision-guided "smart bomb" munitions. This possible loss of production critical to U.S. advanced weapon systems is of particular concern. According to the Defense Department, Magnequench's Valparaiso facility supplies 80 percent of the rare-earth magnets used in the construction of U.S. smart bombs. We also understand that certain magnet production technology is subject to export controls. If true, we believe that the sensitive nature of the Valparaiso facility's equipment may warrant action on your part to prevent its transfer to China . . .

We understand that CFIUS [Committee on Foreign Investment in the United States] approved both the 1995 purchase of Magnequench by a consortium that included two Chinese companies and Magnequench's acquisition of the Valparaiso facility in 2000. However, the potential transfer of these operations to China raises new questions about maintaining both a significant source of domestic production of rare-earth magnets and U.S. technological leadership for these critical production technologies.

We hope that you will direct the appropriate members of your Administration to promptly review this matter. We know you share our concern with any potential loss of critical defense-related production and look forward to working with you on this and other issues.

Sincerely,
Evan Bayh, United States Senator
Peter J. Visclosky, Member of Congress[13]

The White House took no action. "One of the strange things about the Bush administration was that their attitude toward export controls was very similar to the Clinton administration's," says Leitner. "There was no strengthening of the system. In fact, there was a further weakening." As usual, Leitner's Department of Defense stamp of denial for the export of the melt-spinning technology was rejected by the Commerce Department now run by Commerce Secretary Donald Evans, a longtime friend of George W. Bush.[14] The Chinese-owned Magnequench had all the legal clearance it needed. Croat's patent for the magnets, the designs of the production melt spinners, and all the Magnequench computers and intellectual property were on their way to China.

For Peter Leitner, it was all too much to stomach. He did everything he could to bring attention to the dismal state of affairs. He talked to the media openly about his fears. He testified before congressional committees. He even wrote a book titled *Decontrolling Strategic Technology, 1990–1992*.[15] Then in 2007 he retired from the Department of Defense and took a job as the president of a biopharmaceutical company. "A lot of the worst possible initiatives regarding technology transfer and regulatory structure happened under the Bush administration and continued under Obama," Leitner says. "The system is upside down. They're basically controlling old things, but new and emerging technologies are going out the door before the government can move—at the geologic rate it actually moves—to get things under control. So our future is mortgaged. What we're inventing now will be in the world market, creating new capabilities for foreign governments and foreign militaries, and we will live in a continually more dangerous world."

12

The Cowboy

With the world of Wall Street in their rearview mirror, in 2000, Jim and Nina headed to their beautiful property in the St. Francois Mountains to begin their first independent business venture. Jim bought some secondhand excavating machines and dug two spring-fed lakes that he stocked with walleye, largemouth bass, bluegill, and catfish. Jim and Nina hired what Jim called "pot-smoking Amish" to craft a stunning timber-frame cabin secured only with wooden pegs. The building crew was quite different from the employees who'd worked for Jim at Kennedy Capital. "If they ran out of marijuana, they quit working and went home—sometimes for two weeks," says Jim. Jim and Nina funded the venture by cashing out their retirement savings and some Kennedy Capital stock. They called the building Bourbon Lodge after the town of Bourbon, Missouri, and named their company Upland Wings. They designed the lodge for corporate entertainment—exclusive outings for small groups who came to fish and hunt deer, turkey, quail, and pheasant. In the evenings, guests would sit on the front porch, smoking cigars and drinking fine wine while Jim—now a self-taught gourmet chef—prepared fabulous dinners. Nina watched as the fresh air invigorated her husband, much as the southern Missouri mountains transformed Dr. Charles Bunyan Parsons and J. Wyman Jones—both of whom abandoned stressful careers a century and a half before. Jim and Nina were successful lodge keepers, with Nina taking care of the marketing and business side and Jim managing the property and guests. But even with the property booked year-round, it only generated so much income. To make extra cash, Jim put his fleet of earth-moving equipment to use and took jobs for other businesses in the area.

One of his first clients was an aging monstrosity of an iron mine not far from the lodge called Pea Ridge. Jim got wind that the mining company was in trouble with regulators because they weren't complying with environmental standards. He came up with an idea. He could expand his hunting operations by leasing the land around the tailings lake—the area where valueless sand and rubble from the mine ended up—in exchange for doing the remediation work. He approached Tom Gallagher, the mine's senior manager. "Why in the world would you want to lease that tails area out there?" Gallagher asked him. "That's a barren wasteland." Jim told Gallagher he wanted to use it for his bird hunting operation. "I just started laughing at him," says Gallagher. "I said, 'There's no birds out there.'" The tailings pile was a giant heap of rubble with no signs of life in it at all. Jim said, "Oh, but there will be." Then Jim explained that he was starting a "put and take" business, where he would stock the area with pheasant for his lodge guests.

"I said, 'Here's what you're missing, Mr. Kennedy,'" remembers Gallagher. "'We've got a mining operation here. We honestly don't let people on the property with guns.'" Jim promised his parties would carry radios, and the hunt would be called off if the tailings pipeline needed any work. Pea Ridge was in deep financial trouble, and Gallagher was looking for any way to bring in more income. "I said, 'Okay, okay. Let's talk money. What are you proposing?'" Jim said that he'd seen that someone was trying to get vegetation to grow on the tailings pile. He offered to provide the labor and pay for fertilizing and planting twice a year. In exchange, he wanted use of all of the non-mining areas around Pea Ridge—nearly 2,000 acres of land. "That got my attention because I was the dude that was out there in my private truck driving a homemade disc behind it, trying to get stuff to grow on this big tailings area," he says. "I hung it up out there a couple of times because it's just like a big pile of Jell-O." The tailings pile was twenty acres of thirty-foot-deep sand, gravel, and dust. Adding water turned it into slurry and moved it through a pipeline down to a tailings lake. "And guess what? It's like driving a truck over wet concrete. So I'm thinking, If this fool wants to go out there and pay for doing it, that's a big plus for *me*." Gallagher hired Jim on the spot.

On Jim's first tour of the mine, he was immediately struck by the enormity of the engineering project. On the surface was a gargantuan rusting lift that once carried ore from deep below the surface. Gallagher explained that the office building complex and hoist towers were not directly over the ore body. The actual workings of the mine were about a quarter mile away and two thousand feet underground. From there, giant trucks moved ore through enormous tunnels to the hoist. The entire operation made Jim's heart pound with excitement. "It was an engineering masterpiece—nothing short of a small city," he says. The mine had its own water systems, overhead and underground electric utilities, several warehouses, a wood shop, fuel stations, a rail yard, a machine shop, fabrication facilities, crushers, conveyors, and 500,000 square feet of mine processing and office space under one roof. Even though he had never seen it, Jim could almost feel the magnetism of the ore body gripping him from below. "And I had the strangest feeling that I would end up owning it. I didn't know why. I wasn't in a financial position to buy it, and it wasn't for sale. But I had this strange sense that I was going to end up owning it one day."

It turned out that Jim was a natural at fixing a mining company's environmental headaches, and the Pea Ridge tailings pile grew lush and green. He got another reclamation job from the Doe Run Company, which had several lead mining operations in the area. But Pea Ridge was his favorite. Jim became fast friends with the owners and managers, who explained that Pea Ridge was the largest and purest underground iron ore mine in the United States. He learned that the facility had been a robust mine from the 1960s to the 1980s, producing high-grade ore from magnetite deposits—the cleanest source to make pure iron. St. Joseph's Lead and Bethlehem Steel codeveloped the property and later sold it to the Fluor Corporation. When Fluor planned to shutter the operation as global prices tanked, John Wright, one of Fluor's top lieutenants, bought Pea Ridge and kept Tom Gallagher on to manage the operation. The first few years were great, but in the late 1990s, currency exchange rates made the mine uneconomic. Ore imported from Latin America dominated the market, and the demand for iron ore plummeted. "Financially, they were just getting clobbered," Jim says.

By 2001, Pea Ridge was in deep trouble. Modern electric arc furnaces were producing nearly half of America's steel from scrap metal, and a weak dollar made iron ore a no-win situation in the market. "We couldn't afford to pay our electric bill anymore, which ran about $200,000 a month," says Gallagher, "By the time you had the electric bill, payroll, and everything else from insurance to fuel, we were talking big, big money." On top of it, the Environmental Protection Agency and the Mining Safety and Health Administration were cracking down with tighter and tighter regulations. "The government did not help us one bit," says mining engineer Larry Tucker. "The EPA restricted the amount of dust the mine could generate. And we were out in the middle of the *woods*." The electric bill was three months overdue when Ameren, the regional electric company, gave Pea Ridge a thirty-day shutoff notice. Gallagher wrote layoff notices. "It was difficult because I was fifty-five years old, and I was telling all these people I've worked with my entire adult life, 'It's over, fellas. We don't have a job anymore.'"

Then he had to attend to the mine. Without pumps running to extract the million gallons of groundwater seeping in every day, it would slowly begin to fill the tunnels and shafts, eventually destroying all the underground equipment. Gallagher scrambled to save what he could. "We had transformers we had to drain and drag out and a big fleet of mobile equipment." All the structural equipment, like the crusher, stayed where it was. "Then Ameren shut the power off. Boom." Gallagher called Cheryl Seeger—whom he knew loved the mine like he did—to commiserate. "He made the comment about how awful it was seeing Pea Ridge dying under him," Seeger remembers.

Jim was out of town with some buddies when Nina called with the news. "The mine is shut down," Nina told Jim. "The owners are bankrupt. The power is shut off." Jim's response was immediate. "I said, 'Okay. Call the owner, and tell him I'm going to buy the whole thing.'" Nina was not convinced buying a bankrupt mine was a great idea, but she knew that once Jim had an idea, there was almost nothing anyone could do to talk him out of it. They talked it over for hours, and Jim's passion and excitement became contagious. In the short term, the land value alone would be far more than the purchase price. In the long term, prices for iron ore would eventually rebound. To convince Nina, he used the analytical argument that when a useful commodity tanks, it will eventually rebound. Nina loved his enthusiasm about a new adventure, and that was enough to sway her. But the real reason Jim had to have the mine was something he could not logically put into words.

Jim and Nina secured a loan from a local bank, and on December 31, 2001, six days after Jim's thirty-eighth birthday, the Kennedys became owners of what was arguably the world's highest-grade iron ore deposit and three square miles of earth surface in the St. Francois Mountains. The entire package cost just under $800,000. No one was more shocked than Tom Gallagher to find that Jim Kennedy was actually the new owner of Pea Ridge. When Jim first mentioned buying it, Gallagher had thought the idea was preposterous. "I didn't know if he had two pennies to rub together," he says. Before closing the deal, Jim had to assure Missouri regulators that he had a plan for the mining waste on the property. In forty years of operation, Pea Ridge had never had any problems with acid runoff, heavy metals, or processing chemicals getting into rivers or groundwater. The ore was so pure that the resulting waste was a short list of common minerals like silica, calcium, and aluminum. The only two problematic elements from an environmental standpoint—phosphate and sulfur—were locked up in stable minerals, so they didn't leach out and contaminate the watershed. "So for taking on what were some minor environmental liabilities, we ended up with all this real estate," says Jim. "Plus, we knew that there was an ore body underground that one day would be worth something." When Jim brought Mark Haselhorst to see the mine, once again, Mark couldn't believe what Jim had gotten himself into. "It was this vast property of huge hulking rusting buildings from decades ago. I remember Jim saying, 'I'm going to mine this.' And I'm saying, 'How the hell are you going to mine this if a big corporation walked away?'"

Jim's first order of business was convincing Tom Gallagher to stay at Pea Ridge. Jim needed mining expertise, and Gallagher was Pea Ridge's number-one man. There was no longer a working mine to offer Gallagher to manage, but Jim had a plan. "The thing about Jim is, he finds your weakness and exploits it," laughs Gallagher. "Jim knew I love to fish. Fishing is my life." So Jim made Gallagher an offer he couldn't refuse. In Upland Wings' expanding outdoor operations, Tom Gallagher would be the professional fishing guide. The pay was paltry, but soon Gallagher was fishing with some of the biggest names from the St. Louis sports scene who frequented the lodge. More often, his job was to answer his new boss's endless queries about the mine. Gallagher knew a wiser financial move would have been to go to St. Louis to find work, but he couldn't bear the thought of a three-hour round trip commute. Plus, Pea Ridge was home. "I grew up out there," he says. "I got out of school and got married and had my two daughters. I raised them on the mine. Every time I got a carrot dangled to stay at Pea Ridge, I'd grab it."

Jim immediately went to work to generate revenue from the property in order to clear the mortgage payments and perhaps make a little profit. Like the previous owners, Jim sold timber. He sold the tailings gravel to paving companies. Gallagher appreciated Jim's entrepreneurial spirit, but occasionally Jim was a little too creative for Gallagher's taste. "He had a Grateful Dead tribute band out there, and people were camping out in the maintenance shop—what used to be a beautiful shop where you could eat off the floor. They were building campfires and having drum circles." Gallagher was not pleased. "Don't get me wrong. I like the Grateful Dead. It just wasn't the way to treat *my* mine, dammit." Next Jim hired a game biologist to oversee Upland Wings' hunting and fishing operations. "The first thing the game biologist did was bring his bird dogs into the office and turn it into a big dog house," says Gallagher. "And that hurt me greatly. He'd want to go see something, and I'd say, 'Okay, hop in my truck, and we'll go.' And here came a wet lab into the truck with us. It just pissed me off to no end."

Gallagher bit his tongue with the game biologist, but when Jim messed with his maps, Gallagher had to put his foot down. "We had them all labeled, and everything was in a very good order. Jim would just come in and pull shit out and leave it on desks and stuff. That was one of those things that really got under my skin." For the maps that Gallagher wanted to protect, "I took those and put them in a separate map case and said, 'You can do whatever you want with the old mine maps, but this is what I've had approved by the state, and this is what I want to keep sacred. You keep the hell out of these!' And he pretty well did. Finally, I had to come to grips and say to myself, 'Gallagher, it's not your mine anymore. It's their hogs and their corn. If they want to let the hogs in the corn, let them.'"

While many aspects of the new Pea Ridge got on his nerves, Gallagher really liked the mine's new owners. He and Jim got along famously, and he was very impressed with Jim's family. "Nina is a very bright lady, very grounded," he says. "I just thought the world of her and their kids." Gallagher, who also had two daughters, appreciated Jim's parenting style. Nina would go on bike rides over several days, and Jim would bring five-year-old Dravin and three-year-old Shale to the mine. He took the little girls to visit the vast sunflower fields he planted, and they played with the baby ducks he raised in the abandoned mine buildings. Then the three of them would stay over at the lodge until Nina came home. "He'd say, 'I'm playing Daddy,'" remembers Gallagher, who calls Jim "a very good father."

The more time Jim Kennedy spent at Pea Ridge, the more he fell in love with his mine. The more he learned about it, the more it amazed him. To Jim, Pea Ridge was more than a business. It was a symbol—not only of the richness of the earth, but of man's ingenuity in extracting such riches. Pea Ridge harkened back to a time when America was binging on industrialization and forward thinking. Sadly, Jim never put his hands on the actual ore body. Once the electricity was off, going underground was no longer an option. The pumps and hoists no longer ran, and groundwater filled the shafts and tunnels. Outside the main office trailer, however, were two 1,000-pound "nuggets" of pure magnetite—a constant reminder of what was still below.

One of Jim's favorite places on the property was a gigantic warehouse where row upon row of tall green metal cabinets contained the history of Pea Ridge. Inside small drawers like old library card-catalog boxes were core samples excavated by St. Joe's diamond drill back in the 1950s and the successive exploration campaigns that followed. Each core sample included its geographic coordinates and the depth it came from. Jim loved feeling the smooth rock cylinders that unlocked the mystery of Pea Ridge so long ago. The shallowest, youngest cores were sedimentary rock: sandstones and limestone from an ancient shallow sea. Next were volcanic rock, sometimes pink or gray: remnants of volcanic explosions. Deeper cores transitioned into an almost black rhyolite, evidence of pre-Cambrian lava flows. From the very deepest regions, the cores were a homogenous collection of dull black magnetite, a subterranean jewel formed in the final stages of a cooling magma chamber. The core samples were the result of millions of investment dollars by St. Joe's and Bethlehem Steel, and after one of Jim's employees attempted to throw them into a Dumpster, Jim began to worry about protecting them. Jim asked Tom Gallagher what he should do, and Gallagher knew exactly who to call. The following weekend, Cheryl Seeger drove her pickup truck to the mine. She and her husband collected what Seeger decided were the most important samples. "We used the mine maps and tried to get a good representative spread of the mine so people could study it in the future," she says. They picked the best samples from strategic locations in order to reconstruct the geologic history of the mine. Seeger, who was equally heartbroken about the mine's closing, spent several weekends taking core samples from Pea Ridge to the rock sample library at the Missouri State Geological Survey headquarters an hour away in Rolla.

Then Jim began to organize and clean the massive administrative building. There were hundreds of large three-ring binders, folders, and cardboard banker's boxes with documents dating back to the late 1950s. Jim rented a Dumpster and bought several large trashcans with wheels. Before throwing away a box, he would scan the contents. Most of the documents were old personnel files or parts and repair manuals for old mining equipment. Most, he quickly tossed. Nothing was worth keeping until the day he opened up a janitor's closet and found six banker's boxes stored inside. "I set the first box on the trashcan. I took the lid off and looked at the files. The tabs on the folders read 'Rare Earth Elements.' I peeked in a little bit and it looked pretty technical." But Jim had a curious feeling about the word *rare*. "So I took the box and put it back on the shelf and locked the door."

A few months later, Jim again opened the janitor's closet and took the top box to the engineering room. "Over about a week, I literally read through every single document, every page, every word." Unable to understand a lot of the scientific jargon in the papers, most of which dated back to the late 1970s, he noticed a name on some of the reports that was familiar to him. Cheryl Seeger, the geologist he'd met when she came to take the core samples, was an author on many of the scientific papers. He would enlist her help in unraveling the rare earth mystery.

"Jim called me one day and said, 'Hey, what are these rare earths, and what's going on with them?'" says Seeger. She explained the geology and chemistry of the rare earths and how they appeared in massive breccia "pipes" adjacent to the iron ore body. Seeger remembers finding Jim very pleasant, and she enjoyed his enthusiasm for "her" mine. From that very first phone call, Seeger became one of Jim Kennedy's best and most respected friends. "I would literally be reading this stuff, and there would be a vocabulary word that I didn't understand, and I'd call her up and say, 'What does this word mean?'"

To Cheryl, Jim was not just a prospector looking to mine Missouri geology for money. He had a bigger vision. Jim told Seeger that, as a former securities analyst, he had a hard time seeing much of a future for his daughters the way that the country was headed. "And he wanted to personally do what he could to change that," she says. Also, Seeger had a great respect for Jim's ability to understand the difficult geology. "I think for the layman, he's got a very good grasp of the science. And I think he recognizes when he needs to talk to an expert." Jim asked Cheryl what other work the Missouri State Geological Survey had done and made sure he got copies of everything they'd ever published on Pea Ridge. Cheryl and Jim spent countless hours talking about the complicated story of rare earths, and Cheryl introduced Jim to other academics and geologists from the U.S. Geological Survey. "Cheryl was literally my mentor in geology," Jim says. "I earned a merit badge in geology for sure, and she walked me through all of it."

What originally captured Jim's imagination in the old file folders was the word *rare*. But what Jim quickly learned was that despite their name, some of the rare earths were not rare at all. The name was an antiquated chemistry term given to a suite of seventeen elements because they are very difficult to isolate from the minerals they are found in. *Earth* was another old-fashioned geology term: it meant that the metals themselves, like iron, were found in nature as oxides, combined with oxygen atoms. Jim learned from Cheryl that rare earths are actually found everywhere in tiny amounts—just a few parts per million in most mineral deposits. Only in high concentrations are they economically worthwhile. While rare earths are commonly talked about as if they are all the same, each element is distinct in its properties and has different applications. To top it all off, many of the elements have tongue-twister names bestowed on them by the Scandinavians who first discovered them. Trying to remember which element was which with names like praseodymium, neodymium, promethium, samarium, and gadolinium was, for the layperson, nearly impossible.

Most economic deposits are primarily composed of rare earths with lower atomic numbers, which are then classified as light rare earth elements (LREEs), because they have a lower atomic weight. Other geologic deposits contain rare earths with higher atomic numbers, referred to as heavy rare earth elements (HREEs). In general, lights are cheap and abundant; heavies are expensive and increasingly rare. Each individual oxide or metal processed from that oxide sells for a different price, based on its purity. There were no formal exchanges, and the "markets" were obscure multileveled microcosms that were next to impossible to understand. The entire scientific and commodity realm of rare earths was mind-numbing.

For Jim, that just made it all the more interesting. He was a quick study, and while he had to ruminate on every word he read, his brain remembered everything he forced on it. What he soon recognized was not only did Pea Ridge have an immense body of chemical-grade iron ore but his mine held tremendously rich rare earth deposits. A 1990 U.S. Bureau of Mines report estimated that the Pea Ridge breccia pipes held 600,000 metric tons of rare earths. The average grade was 12 percent, and the pipes held a large amount of the more pricey heavy elements. Kennedy's experience in finance helped him understand the economic significance of this, but what he soon learned about the applications of rare earth elements in advanced U.S. weapons systems also helped him grasp his mine's value for America's national security.

Jim spent every waking moment working out a way to bring the mine back to life. Sometimes, says Tom Gallagher, his ideas were terrific. "Golly, does he come up with all kinds of good research. He would start reading and gathering information, and he retained the information. It was really something to watch. He just grew every day on the subject." Other times, Gallagher would spend hours trying to convince Jim that his idea wouldn't work. "He lacked the practical application of it. He took everything as gospel and thought that all things will work in all places. It just doesn't work that way."

After hundreds of hours of research, Jim was convinced that Pea Ridge could be brought back to life. And if he could prove it, he could sell the whole mining operation for a nice profit and keep the land around it for Upland Wings' recreational operations. Gallagher wasn't convinced. "Here's the deal," Gallagher told his new boss. "There's a million gallons of water a day dumping into that deep hole. You cannot restart the existing pumps in the mine." Gallagher explained that the motors and all their components were underwater and not designed to be underwater. "Jim told me, 'Oh yeah, yeah. We'll put some pumps in, and we'll hang them below the cages, and we'll keep dropping the cages.' I said, 'Jim, the hoist ropes have been hanging under water for four years or better.'" Jim told Gallagher that it wouldn't be a problem. "I'm thinking, 'Jim, you're not thinking straight. It *will* be a problem.'" Finally, Gallagher told Jim that before he started talking about reopening the mine, he needed to hire some serious engineering companies.

While Jim studied every aspect of mining engineering and economics, he became increasingly transfixed by his mine's rare earth elements. He was desperate to incorporate them into the engineering project, but he didn't know how. "Everyone kept telling me that I needed to keep the assessment simple," says Jim "The experts and engineers said, 'Focus on the big stuff. This is an iron ore mine. Don't complicate it . . . Sideline issues do not add value. Just show that Pea Ridge is economically viable.'" But Jim couldn't get the mysterious files out of his head. That winter, late in the evening in the unheated and therefore freezing engineering room, he pored over documents for hours upon hours.

He soon learned his ore deposit contained a phosphate mineral called apatite that was high in rare earth concentrations. "The reports downplayed apatite as a potential resource at Pea Ridge," says Jim, "mostly because the company was not in a financial position to pursue it." Apatite is a byproduct of iron mining, and the tailings lake held nearly forty years of accumulated apatite. Jim made some quick calculations; the tailings lake could hold over $1 billion in recoverable rare earths—from just the surface—and another $3 billion underground.[1]

"I tried to figure out how, when the mine reopens, we could start developing the rare earth deposits. The questions were: Who's the end user? Who's the customer?" A few dozen Internet searches jarred Jim into understanding the economic reality of the rare earth market. "I quickly discovered that China alone mined rare earth ore, processed it into metal, fabricated it into magnets, alloys, and components, and then sold it to end users. There was no value chain in the United States at all." Molycorp, the one mining company in the United States solely dedicated to rare earths, had closed in 2002. There were still a few refineries left in Japan, but "the Chinese were basically killing off all the Japanese competition, too." To make matters worse, the Pea Ridge rare earth minerals also contained the radioactive element thorium, and Japan had strict import restrictions against thorium at even the lowest levels. So China would be the only potential customer.

"By 2008, I understood this stuff very well, much better than most people," Jim says. He talked at length with U.S. Geological Survey geologist and rare earth expert Bradley Van Gosen, who informed him that there were many potential rare earth sources in the United States. But ultimately, China was the only market. China controlled the value chain. China was the ultimate consumer. China set the price. "I began to realize that for China, controlling rare earths was not about making money," Jim says. "It was about capturing industry, technology, jobs, and intellectual property." The U.S. manufacturing and defense industry was intractably tied to the Chinese monopoly. He may not have been a member of the U.S. military any longer, but patriotism still coursed through Jim's veins. There was no way that he would sell *his* rare earths to China.

At the time, rare earths were a mere $3 billion industry—less than 1 percent of the iron and steel industry—so they were ignored by commodity traders on Wall Street and overlooked by procurement analysts inside the Pentagon. However, they were critical components of a consumer and military market worth $5 trillion a year. Once they secured a monopoly, China would control not only rare earths but every single consumer and military product that needed them. Jim became obsessed. He believed that the situation was a disaster for America—and that, just maybe, he could have a role in fixing it. "It seemed no one in Washington, D.C., recognized the threat. So that's when I started reaching out to members of Congress and the Department of Defense and alerting them to the reality that this stuff is used in dozens of weapons systems."

For months the problem tormented him. Then one day he came up with a solution.

13

The Engineer

Although he calls himself a typical nerd, with a subscription to *Popular Science* and *Omni* magazine, over the years John Kutsch has developed a public acumen that most of his fellow engineers notably lack. It began in the Chicago suburb of Glenview in 1967, when John arrived "as a mistake," he says, eight years after his next youngest sibling, in December 1967. His parents—in their mid-forties when he was born—were as old as some of his friends' grandparents. By the time John was ten, he was the lone kid in a big house, and "got to do pretty much whatever the hell I wanted." His father was a World War II hero and a well-respected hospital administrator whom young John held in very high esteem. His mother was diagnosed with severe multiple sclerosis at the same time she found out she was pregnant with John, her eighth child. "Even though she was crippled and in a wheelchair most all my life, she was still the toughest lady in the world," he says. She was also well educated, a rare female college graduate in the 1940s. By the time John was in school, his parents "were distinctly over the whole 'raising kids' thing," he says. "Checking report cards and college planning were long abandoned." John spent his free time building things and taking them apart again. He wasn't much of a student at his prep school—the Loyola Academy in nearby Wilmette. What set him apart from other machine-loving nerds was that he acted in school plays and volunteered for every outreach program he could through his Jesuit school. He visited elderly folks, worked in a soup kitchen, and helped out with underprivileged kids.

John pulled off a very high SAT score, but didn't feel the urge to follow his friends to Ivy League schools back east. Without a push from his parents to do otherwise, he chose the same second-tier school that one of his brothers graduated from. The Loyola Academy college counselor called him out in the hallway. "She said, 'What are you thinking? You're going to shame us by going to Southern Illinois University. Can't you apply anywhere else?'"

As soon as he arrived in what he calls the "cultural desert" of Carbondale, Illinois, John plotted his exit strategy. He took an enormous course load, intending to graduate with an engineering design degree as soon as possible. He was up at seven in the morning and in school until ten. He served as vice president of his fraternity, Tau Kappa Epsilon. "It got me used to trying to manage a bunch of cats who wouldn't take direction," he says. Even better for a guy from a white-collar suburb, the fraternity was incredibly diverse. "I had roommates who paid for college by de-tasseling corn. I had roommates who paid for college working on their dad's farms. The TKE house was racially mixed, and its members studied a wide variety of majors. I would never have that if I went to Boston College or the Rhode Island School of Design. I would have been surrounded by a bunch of white suburban prep school kids just like me."

John soon found that the drawback of college was that his classes didn't give him any practical engineering experience. On summer break after his junior year, he went to find some. "I looked all over for an internship, and the miracle was, I found Cesaroni Design Associates just a few blocks from my parents' house." John took his portfolio with his classwork and walked into the building's front office. I'd like to have a job, he told the owner, Bill Cesaroni, with a big confident grin. No pay necessary. Cesaroni took John's book, flipped through it, and scowled.

"Ugh," he grimaced. "This is terrible!"

He turned a page.

"Jesus Christ, look at this!" he grumbled.

Cesaroni flipped to the next page.

"Goddammit! This is what they teach you?"

John gulped and made a mental note to call the welding company he'd worked for the previous summer. Then Cesaroni said, "This is such an appalling portfolio of work that I've got to hire you so you can learn something."

Cesaroni turned out to be a great businessman and an excellent designer who made good-looking, sturdy machines for all sort of industries. John credits Cesaroni for changing his young life and giving him some really stellar things to put on his résumé.

When John graduated after four years of college, it was 1990. The Berlin Wall had fallen the summer before. "The Cold War was over, and there were no engineering jobs to be had on planet Earth." He and his fraternity brothers sat on the porch of the TKE house, drinking beer and commiserating over nightly NPR reports of the "worst job market in a decade." He moved back into his parents' house, did some construction work, drank a lot of beer, and drove a hearse he bought for $200 from a friend's brother. To make up for his shoddy degree from a low-ranked school, he started an MBA program at Loyola University in Chicago and waited for his luck to change.

A year and a half passed before John got word about an opening at Baxter International Inc., a giant medical equipment company with 50,000 employees. It had been five years since Baxter took an outside hire. John was desperate for the job.

"What do you think you can bring to Baxter?" the interviewing manager asked.

John might have been out drinking after his construction job every night, but he still kept up with the latest technology news.

"Let me tell you, I know what the future is!" he said excitedly. "The future is this stuff called stereolithography." John leaned forward in his chair for emphasis. "It's where you can take a computer-designed part and divide it up into little tiny slices and grow it on liquid polymer with lasers. You can actually make a full working part straight out of a machine—like printing a part!"

"Oh yeah?" the manager said, laughing. "And do you remember reading about where that machine was?"

"Uh . . . well . . . I read it in *Plastics Engineering*."

"Yeah, but do you remember *where* that machine was?" the manager asked again.

"Uh . . . no . . ."

"You're standing one floor above it."

John was—uncharacteristically—speechless. It was 1992, and the twenty-five-year-old was about to see the world's first 3D printer in action.

3D Systems—the company that patented 3D printing in 1986—sold one of their very first stereolithography machines to Baxter, and Baxter was helping develop it. "You can't imagine what it was like actually seeing a part being grown out of liquid plastic." To the kid who grew up on *Star Trek*, it looked like a teleporter. "You could design a part on a computer screen and essentially hit print. Then a few hours later, the part would be in your hand," says John. "It's still unbelievable to see it today."

He got the job in Baxter's brand-new Ambulatory Infusion Business division. It was a techno geek's dream come true. He had access to the latest software and the most expensive computers from Sun Microsystems and Silicon Graphics. "I had a $40,000 color printer, just for me," he says. "It was like going to work at the space station every day." *Need something expensive and state of the art? Ring up procurement and just ask.* He even had his very own office with a great big window.

John designed intravenous devices for the first year, then transferred to Baxter's Advance Engineering and Design Center. The design possibilities were almost limitless because the team created things not only for Baxter but for outside companies as well. John designed farm equipment for Caterpillar, cell phones for Motorola, and landing gear for Boeing. He loved it all. "It was an unbelievably fantastic opportunity," he says. But he wasn't completely fulfilled. "Eventually, I got a bug in my ass. I was a typical guy in my twenties. I knew everything there was to know. So I went and started my own design engineering firm." It was 1995.

John Kutsch became the twenty-eight-year-old president of the newly formed Whole World LLC. By the late 1990s John had as many as fourteen engineers working on any one project. He got contracts with Motorola to work on the very first smartphone. He worked on diagnostic machines for Abbott Laboratories and designed heavy equipment for Toyota. His team built X-ray equipment for a dental company and office supplies for ACCO Brands Corporation. He even designed airplane toilets for EnviroVac. A company called Medline Industries became his biggest client, contracting everything from a simple bucket to advanced surgical tools.

Although his client base was stellar, owning a design firm was harder than he'd thought. He wasn't able to compensate his engineers like they needed to be, and he felt terrible when he had to let them go. But he knew they'd be much better off financially going back to 3M and the other Fortune 500 firms they'd come from. John kept a partner of sorts in design engineer Vince Lackowski, and Whole World LLC did quite well with just the two of them and some freelance help into the 2000s.

There was, however, a sea change across the American Midwest. "I was losing clients left and right. One client after another closed up shop and left for China or India." To John, it was all part of a growing strategy that he believed would debilitate the U.S. economy. "In the mid-nineties, when I was getting my MBA, they started yapping, 'Oh, it's the *idea* economy. We're going to let the workshops of the world take over and make stuff. Dirty, dull, dangerous manufacturing is a thing of the past for America! Get rid of the bricks and mortar and just stick with the sales and design.'" John believed that it was a doomed strategy that would never work long-term. "What kind of quality of life is this country going to have if we don't make stuff?" he says.

In 2002 John ran for U.S. Congress for the Sixteenth District in Illinois. He lost. He tried again in 2004 and lost. "My political aspirations were very self-serving. I needed my clients to be in the United States in order for me to have clients at all." It also served his patriotic sensibilities. "I always went out of my way to specify American devices and American components in my designs whenever I could. I would fight like crazy to get stuff cast and built and injection molded in the USA."

One day in 2005, one of John's clients—a renewable energy company—came to him looking for help. There was a problem with something they were trying to build, and they needed a substitute for one of the metals. John was not a chemist, but he had a pretty good knowledge of how metals worked in electronics. He and his client scrolled through the periodic table. *Will cobalt work? No. Will tin work? No. Will thorium work? Hmm . . . Not sure.* All John knew about thorium was that it was mildly radioactive and formerly used in Coleman lanterns to enhance the brightness of the gas flame. His client asked John to look into it.

John googled thorium and happened upon a website run by a guy named Kirk Sorensen. There was a chat room with "all these nerds talking about something called a thorium molten salt reactor." There were pictures of it at its home at the Oak Ridge National Lab. There were reports about it.

"It wasn't just a bunch of propeller heads postulating about a new 'concept' commercial nuclear reactor that could never come to fruition. It was real."

John had always had a love for nuclear—more of a pet love than anything he pursued. Growing up, he had a friend whose father was a nuclear engineer and told him about lots of new nuclear design possibilities. Another friend's father was the head structural engineer at Fermi National Laboratory and brought John to visit the lab's giant particle accelerator. Learning early on what was possible made John completely unimpressed by the current reality of commercial nuclear energy. "It always seemed so primitive to essentially just stick some really radioactive rods *really* close to each other until they get super excited and boil water." John knew that what were now called "advanced" reactors were just the same old reactors built with "tons of safety systems and bells and whistles," all to keep the core from melting down or the containment structure from exploding. "I had it in my head, there's got to be a better way."

John looked at the site for over an hour and marveled at the TMSR design. "I instantaneously got what this thing was about. I knew if they were able to do it once, it could be done again—and it could be commercialized." To John, climate change is a critical issue for the future of the planet, and having a robust nuclear energy industry would certainly help curb carbon emissions. He wanted to get this done. Unfortunately, his real-world sensibilities didn't mesh with the other geeks on the forum. "Has anyone tried to get this thing restarted?" John posted. "Is there any of this stuff going on at universities? Does Oak Ridge still work on this stuff?"

"They'd be like, 'Well, maybe Oak Ridge is still working with salt . . . Speaking of salt, I've got this new idea . . . let's talk about my salt idea.'" John would try to keep them on topic and get someone to commit to working on bringing the TMSR concept back to life. "But it was like talking to attention deficit disorder kids. No one ever listened." He occasionally jumped back on Sorensen's chat room and pushed for a concrete plan. Then he'd get discouraged and log out again.

A few years went by, and in 2009 John still couldn't get his mind off the thorium reactor. New and safe nuclear power, in both a political sense and an environmental sense, was too important. "You will not provide energy for 12 billion people [an estimate of the global population in 2100] with solar, coal, wind, or gas," he says. One pound of thorium used in a molten salt reactor is equal to as much as 400,000 pounds of coal; same goes for oil and gas. One unit of thorium used in a TSMR is worth several hundred times that of solid uranium in a light-water reactor—a factor of the "energy density" of fuel. "All the ways to produce energy require resources dug from the ground and manufactured with a lot of energy and cost in labor, land, and lives," John says. "Since that's the case, we *must* choose the most energy-dense, safest, and cleanest energy source—which is nuclear."

John decided to move forward on his own. With the proceeds from a building he sold, he could afford to dedicate twenty hours a week to work on thorium nuclear projects. He called his endeavor the Thorium Energy Alliance, and began to postulate what his new alliance might do. John and his partner Vince called up some architects and chemical engineers to work up some numbers. A simple, unfueled test reactor simulator would cost about $94 million to build. It was 2009, and the world's economy was in the toilet. John knew that no single university, especially in these tight financial times, could afford it. John looked at situations where universities partnered with the National Science Foundation on projects like giant telescopes and came up with a similar plan. He would get every nuclear school to come on board. "I was very provincial," he says. "I wanted to do it at the University of Illinois." The university was already working on a license to build a teaching reactor in the Illinois countryside.

Another part of John's strategy was for the schools to bring the project political clout. "Universities spend huge amounts of money lobbying the government for handouts. *Let's build the new nanotechnology lab. Let's build the new bio lab.* These guys are experts at wringing money out of the government. I figured, if I can get ten to twenty senators and thirty to a hundred congressmen on board, that's my political base." Then John would get the universities' lobbyists to press the National Science Foundation, the Department of Energy, and the Department of Defense for funding. "I thought that would be a decent way to go about skinning that cat."

In the fall of 2009, John put together his first Thorium Energy Alliance conference. "I was going to turn my initial battalion of nerds into little evangelists. I would send them to congressional staffers to talk about the thorium molten salt reactor and how it's a real reactor and it's not risky." In his frantic search for meeting content, he asked a friend to recommend speakers. "I got a list: Jim Hendricks, a USGS geologist; Jack Lifton, an analyst for precious metals and mining; and Jim Kennedy. I was driving to my farm and calling down the list." When he reached Jim, he "sounded less than thrilled about the invitation. He was like, 'Uhhh . . . Okay, I'm going to be in D.C. then anyway. I'll stop in.' So with that ringing enthusiasm, I had a guy who was gonna talk about rare earths. I didn't even get why. I was just like, 'Thank god, I got another speaker.'"

14

Pay Dirt

When John Kutsch called to invite Jim Kennedy to his thorium meeting in the fall of 2009, Jim sounded a bit distracted because his life was suddenly in overdrive. And it was all thanks to the People's Republic of China. The Asian giant was growing wealthier by the second, and had endless cash to finance a mega growth spurt. Massive skyscrapers were springing up, extending and transforming ancient villages into major cities and industrial centers like tentacles of a giant octopus. Bold steel-framed structures replaced bamboo, concrete, and block buildings. Modern new cities arrived seemingly overnight. There was very little scrap steel, so China was wholly dependent on virgin iron ore and integrated steel production. The building boom created global shortages in iron ore, copper, and just about everything else needed for construction. Western mining industries—starved of investment after years in the doldrums—could not keep up with demand. Between 2004 through 2006, iron ore prices rose nearly 300 percent. Speculation that the boom would continue sent prices through the roof, creating what economists call a "super cycle" in the commodities market. Lots of new production in both mining and recycling came online to supply China—so much so that by 2003, China surpassed Japan as the world's biggest importer of iron ore and just about every other construction commodity.

In 2001, when Jim and Nina bought the pristine magnetite deposit at Pea Ridge for under $1 million, the price of iron had fallen to about $13 a ton. Jim expected a rebound one day, likely in the far-off future. Instead, five years later, the price of iron ore exploded to nearly $100 a ton. The Pea Ridge iron reserves were now worth over $1 billion. The thought was mindboggling. Jim and Nina decided that it was the perfect time to cash out on the mining and mineral rights to Pea Ridge and keep the surface property for their lodging and hunting businesses. Unfortunately—even with ore prices sky-high—they soon found that selling Pea Ridge wasn't as easy as they hoped. A few potential buyers came knocking. They told Jim the mine wasn't economically feasible and made dismal offers—barely enough to cover what Jim and Nina had put into the property.

Refusing to believe that Pea Ridge couldn't make a profit, Jim decided he would bring the mine back online himself. With his twenty-four college credit-hours in geology and his new love of all things engineering, Jim devoted himself entirely to the mine. When Tom Gallagher wasn't out fishing, Jim picked his brain on all aspects of the ore body. He hired former Pea Ridge engineer Larry Tucker as a consultant, tossing out ideas that Tucker analyzed and grumbled about. When Tucker gave Jim a no-go, he would just go back to the drawing board and come at Tucker with another idea, then another.

To the old Pea Ridge miners, Jim was more an idea man than a reality guy. They appreciated his passion, but he would never fully understand *their* mine or be connected to Pea Ridge like they were. He couldn't, said Gallagher, because he didn't have the history. The one person who believed that Pea Ridge ran through Jim's veins as thickly as it did the others' was Cheryl Seeger, who Jim now had on speed dial. Seeger often came to visit "her" mine and sat with Jim for hours, going over the extraordinary geology of Pea Ridge—a subject neither of them ever tired of.

To bring the mine online and find investors, Jim would need to hire some serious and well-respected mining companies. He found an Australian company that specialized in deep ore bodies and invited their managing director, Ben Auld, to come to have a look. Auld was delighted. He already knew about Pea Ridge because the mine was featured in his college textbooks at the West Australia School of Mines. At the time Pea Ridge was built, it was the leading edge in underground mines. "I said, 'Absolutely. I would love to be involved,'" remembers Auld. He knew he would really enjoy the challenge. What he didn't know was how much he would enjoy working with the mine's owner. "We Australians are very direct in our conversation and what we're thinking," he says. "Jim is exactly that if not more so. We got along like a house on fire."

A few days after he arrived, Auld rode shotgun in Jim's SUV back to the lodge where Jim put up his visiting experts. Jim's phone rang. The call was obviously about some financial matter.

"He's driving the car, yelling into his phone, and he goes, 'Are you serious? You might as well put a gun to my head and just rob me!' In between swearing and yelling, he hits the phone on the steering wheel. Then he hangs up. He turns to me, and he just smiles. And he says, 'Watch, he'll call back.'"

The person called back. "Jim's swearing at him, 'You're just a robber! Pull the trigger!' And he hangs up again." Clearly delighted, Jim turned to Auld and said, "Ah . . . This is great. I love it."

"I'm like, 'Holy shit, this is mad.'" Auld couldn't help but be drawn in by Jim's personality. "He's just a high-energy, high-passion guy. At the same time, it's all a bit of a show. He *is* a bit of a showman."

Auld quickly realized with the market conditions and the high grade of magnetite, the mine could be profitable once again. Auld was especially impressed with Jim's enormous plan. "It was the right thing to do for a long-term vision," Auld says. "That's something that's really rare for a mine owner. They typically get caught up in capital wealth and cash flow. Jim was more interested in doing the right thing about the deposit and value-adding all the way down." Auld was also taken with Jim's capacity to have a handle on all aspects of the project. "Far beyond understanding the geology of the mine, he had this inordinate amount of brainpower to consider all the subsets and aspects," he says. "I met Cheryl Seeger, and obviously he had pretty much drained her brain of everything she knew about Pea Ridge. He knew not only about the geology of the operation, but also what the geology meant." Auld joked at the time that if he had the authority, he would have given Jim an honorary degree in geology. "When you have someone like Jim who looks at it in terms of not only years but decades, that's just fantastic." Auld thought that the way that Jim put all the pieces together was a true feat of engineering. "So I said, All right. This is a man on a mission. We'll do whatever we can to help get the right answers to make it happen."

Now Jim needed a way to pay for the "pre-feasibility" studies that could run in the millions of dollars. He had an idea. Back in 2003, before he ever thought about reopening the mine, Jim was gung-ho about expanding his hunting business. There was little level ground on his three-square-mile property, but downhill from the mine were two flat areas spanning several acres. Lush vegetation grew in both expanses—giant cattails up to ten feet tall. Jim wanted to cut some lanes through the flatland so it would be accessible to the hunting parties. To clear a road, he drove his five-thousand-pound Massey Ferguson tractor down to the lower area with a brush hog (a rotary mower) attached on the back. When he hit the first expanse, the ground was firm and solid under the enormous tractor. In the cab, Jim was just higher than the cattails, and couldn't see the vegetation giving way to the brush hog behind him. Jim made a U-turn to give the trail a second pass. Then he went back over the spot he'd already cleared. The ground was jet-black, and had turned from solid to liquid; water was seeping up to the surface.

On the third time over the same area, the tractor just dropped through like a heavy kid on thin ice. Jim jumped off the tractor, which was now stuck two feet deep in the muck, grabbed a bit of the black material, and rubbed it between his fingers. It was heavy, really heavy. Jim knew that the only thing it could be was pure, processed magnetite. Jim called a tractor-trailer towing company and watched the tractor emerge from what looked like a bottomless black pit.

Jim asked the old-timers—Tom Gallagher and Larry Tucker—where the mystery magnetite came from. They told him that from 1958 through 1959, Bethlehem Steel, even without strict environmental laws, had been a good corporate citizen. To avoid waste from the processing plant going downriver, the company dug two "settling ponds" at the bottom of the valley. Over fifty years, all the drainage from the processing went into storm drains. The storm drains dumped into the settling ponds. It wasn't tailings. It was pure, finished magnetite product.

Jim went to work reading dozens, even hundreds of documents on processing iron ore. On paper he designed a benefaction (ore processing) operation. He ran his ideas past Tucker, Gallagher, and other Pea Ridge veterans. Moving the sludge to process it on the building site would be a costly and messy endeavor. Instead, Jim decided to build the plant right there in the settling pond.

Jim went home and convinced Nina they should cash out the rest of their Kennedy Capital stock to fund his endeavor, which he calculated, in the end, would generate enough income to pay for the engineering studies on Pea Ridge. Then Jim hired a handful of former Pea Ridge workers to help. Those working with Jim remember his dedication. They bought two 20-by-10-foot boiler tanks made of three-quarter-inch steel, cut the tops off, and opened them up. For an entire week that summer, Jim stood sweltering inside the steel tank, sandblasting the rust out. He found old pumps and magnetic drum separators to pull out the magnetic particles. Jim became friends with an old metal scrapper who built Jim stuff to spec out of things lying around his junkyard.

With this five-man crew, Jim worked seven days a week, sunup to sundown. One day, when he was down a welder, Jim asked one of the old-timers for on-the-spot welding lessons. He worked with the welders the rest of the day. At night at the lodge where he stayed most nights, to avoid the hour and a half commute back to St. Louis, he read about mining innovations and the metallurgy of pig iron. A few times, Jim's entire body and mind would give way to the fatigue, but the crew says Jim was on a nonstop adrenaline high no matter what obstacles came up.

One problem after another plagued the operation. Weeks went by. June gave way to September 2007, and the leaves in the St. Francois Mountains turned red and gold. Finally, one day, the machine was ready. They loaded the black sludge in, sloshed it in clean water, and pushed it through several screens. The magnetite ore was so dense that the slightly lighter silica and hematite floated out. The first batch looked beautiful, black and pure—tiny magnetic particles lined up to the pull of Jim's magnet. The crew held their breath as Jim sent a sample off to a lab in Minnesota.

What he hoped to find was that they'd reached "chemical purity." Magnetite, being a combination of iron and oxygen with three iron atoms for every four oxygen atoms, is 72 percent iron by weight and 28 percent oxygen by weight. To sell the magnetite as chemically pure, Jim needed his sample to be at least 69 percent iron by weight, and 96 percent magnetite (Fe_3O_4). When the sample came back a week later, it was 2 percent shy of chemical purity. Jim's workers could see he was beyond stressed. They knew he was basically out of money. Back home in St. Louis, with two small daughters and Jim rarely home, Nina tried to remain calm, remembering their friend Rich Todaro. As the full-time bookkeeper for all of Pea Ridge and Upland Wings operations, she had a front-row seat to their dire financial straits.

Week after week, Jim pulled random samples and sent them to the lab. Finally, one of the workers took a pipe and stuck it high up on the tank. The very fine hematite floated to the top and poured right out through the pipe. What was left was unbelievably clean magnetite. The lab test came back the next week—better than chemical purity.

As soon as he could, Jim found a customer to buy his product—a Florida water purification company. The first year was a success, and he had double the orders for the second year. Now there was money to start the engineering studies.

With the newly found income, Jim hired Ben Auld's Australian company and another Brazilian engineering firm. Jim was in heaven. When the engineers were on site, he was there too. During the day he followed them around, asking questions and getting to know the mechanics of the mine with the utmost intimacy. Auld remembers that Jim offered endless ideas—sometimes grandiose schemes that were politely refuted, sometimes brilliantly thought-out ideas that stunned all the engineers.

All the while, Jim and Nina searched for the most important piece of the puzzle—the money to make it all happen. Auld says that the final project would cost hundreds of millions to $1 billion—possibly more. Jim and Nina had interest from Chinese investors who agreed to put the money up in exchange for exclusive rights to market the ore. But months went by, and the Chinese still hadn't committed to financing the entire project. Finally, when they got wind that the real estate for the smelter was about to be hijacked by another developer, the Chinese stepped up with enough money to jump-start the project. The major funding would follow.

It looked as if Jim Kennedy's dream was about to come true.

15

The Vision

With Pea Ridge financing in the works in the spring of 2007, Jim Kennedy opened up a map on his kitchen table one morning, placed a string on the Pea Ridge mine, and moved it in an arc to the east along the Mississippi River. Then he highlighted all of the riverfront areas that fell within the smallest possible arc. Part of Jim's engineering plan was to transport the iron ore from the mine to the Mississippi River via a fifty-mile pipeline. The savings in transport costs were what would make the mine ultimately profitable. Once by the river, he would smelt the ore into pig iron and ship it off on river barges, so he was looking for the perfect location for the end of the pipeline along the Mississippi. The place he would choose had to have river access, rail lines, natural gas pipelines, industrial electric service, and interstate road access. With two possible suspects, Jim and Nina hopped into their '66 Mustang, put the top down, and headed south from St. Louis, hugging the western bank of the Mississippi.

They first came upon the town of Herculaneum, an industrial port city established by Moses Austin in 1808. Here, in 1892, St. Joseph Lead built what became the largest lead smelter in the world, bringing a century of working-class prosperity to the town. In 1994 Doe Run Company bought all the lead mines and the smelting operation from the Fluor Corporation, but with the deal came a thick catalog of future liabilities. St. Joe's smelter had been mostly unregulated for the bulk of its history, and the lead-sulfide ore produced acid rain and soot that coated the town day and night. The resulting lead permeated soil, homes, and waterways. Doe Run paid out settlements in the tens of millions, and the EPA relentlessly pursued the company on all fronts.

Against the backdrop of giant sandstone cliffs, a 550-foot smokestack stood guard over rusting beige warehouses. They couldn't enter the site, but Jim and Nina drove around the perimeter to get a peek. There were lots of positives, but there was no way to get around the enormous environmental liabilities, so Herculaneum was not a viable option.

Jim and Nina continued south. In just ten minutes, they drove into Crystal City. They found the town to be quaint and charming, with brick and stone houses on one side of the main street and smaller wooden houses on the other. Like Herculaneum, Crystal City also has its roots in Missouri geology. But in contrast to its northern neighbor, the prosperity of Crystal City was born of some of the cleanest and purest mineral deposits in the world. In the early Paleozoic era (around 500 million years ago), a shallow sea covered a topographic depression known as the Illinois Basin. Over millions of years, rain and wind wore down quartz-rich mountains to the north. Rock particles tumbled down rivers and creeks, creating beaches along the sea. Waves churned and polished the rubble for eons until only quartz—the most stable mineral on the earth's surface—remained. As climate changed and beaches waned, deep layers of sand—in some places six hundred feet thick—compacted and turned to stone. Over the next few hundred million years, the moving and shifting of tectonic plates buckled the brittle sandstone and pushed it to the surface, to be eroded once again by wind and rain. This time, the quartz sand found its way into the western tributaries of the great Mississippi River. And it was there that a geologist named Forrest Sheppard stumbled upon the tiny frosted grains, white as refined sugar. Sheppard collected a sample near the mouth of Plattin Creek in Jefferson County, Missouri, and sent it to England. The sand was 97 percent pure quartz—some of the purest sand ever discovered, and perfect for making glass.

Captain Ebenezer B. Ward from Michigan bought the claim and founded the American Plate Glass Company in 1871.[1] He called the town New Detroit. The company changed hands in 1876, and the Crystal Plate Glass Company expanded operations, with four gas-fired furnaces and a large plate-glass operation. Glass was poured onto large square tables, ground with sand, smoothed with an abrasive powder made from garnet-rich rocks, and then polished with iron oxide. In 1895 the Pittsburgh Plate

Glass Company, later known as PPG Industries, acquired the factory, the company town, and all of its holdings.[2] Residents of the young town, looking for their own identity, christened the town Crystal City. It was a company town until 1906, when private citizens were allowed to purchase lots and build homes. By the 1940s the glassworks covered eighty-five acres, with about twenty acres under one roof. Two thousand residents of Crystal City and nearby Festus worked at PPG, a well-planned development that resembled a college campus more than an industrial plant. In 1985 industry consolidation in the manufacture of automobile glass eventually forced the Crystal City plant to shut down. By 1991 the entire complex was completely gone except two remaining structures—the old hospital and the old company headquarters, both used for storage.

Jim and Nina walked around the fence line of the expansive grounds. Jim's heart raced. "The PPG site was a blank slate," he says. "Almost all of the buildings had been removed long ago. PPG had finished an exhaustive state-monitored remediation program, so the liabilities were well-known and manageable." The site had direct access to the Mississippi, Union Pacific, and Burlington Northern rail lines and connections to electric and gas lines. It was perfect.

Before Jim contacted PPG's Pittsburgh headquarters, he made an appointment with Crystal City mayor Tom Schilly and a few of the mayor's advisers. Jim took a deep breath, pitched his plan, and crossed his fingers. The city had been in decline since the closure of PPG, and the council was dying for an economic boost. "I was hoping they would support such a large industrial project in their backyard," Jim says. The mayor was very receptive, and Jim liked him instantly.

"When Jim brought the plan forward, I have to be honest with you, I thought it was a pipe dream," says Schilly. "But as persistent as Jim is, one thing led to another. It seemed like the right hand knew what the left hand was doing." Schilly, who drives thirty miles to his job as a welding supervisor every weekday before heading to city hall, was cautiously optimistic. "I'm so sick and tired of sending our jobs overseas," he says. "All I wanted to do is just bring a good, clean industry back to Crystal City. Jim's plan was to be a good clean industry—no smokestacks, no coal-burning furnaces. I thought it was a great idea." Schilly called a meeting with the city attorney and city council. The fully developed operation would result in 750 new, high-paying jobs. The vote in favor of the project was unanimous.

There were at least three other groups pursuing the PPG property. To secure it, Jim needed to bring the city in as a partner. Schilly remembers Jim's plan: Jim's company, Upland Wings, would provide the city with funds to purchase the property through a prepaid hundred-year lease. This would give the city some control over the project rollout. Jim was obligated to be under development within five years. To keep the deal confidential, the mayor and city council members signed nondisclosure agreements. Everything was falling into place. Jim and Schilly were on top of the world.

Then all hell broke loose.

Within days of the deal announcement, a loud uprising by a group called Concerned Citizens of Crystal City railed against the "smelter" project. Jim, Nina, the mayor, and city council members were blindsided by the vitriol. Protest signs saying things like "Stop the Smelter!" went up all over town. A website sprang up, with thousands of anonymous posts emphatically deriding the project. Protesters came to city council meetings wearing gas masks. "I was surprised, because all we were trying to do as the city council was bring forth jobs, industry, and revenue—not only for the city but for schools and the surrounding area," Schilly says. "Like I told the wife, 'I don't know who the people are, but I turned over somebody's apple cart that wanted the PPG property, and they didn't get it." Concerned Citizens of Crystal City boasted two hundred members, but Mayor Schilly says that there were really only a handful of locals involved; the rest were from out of town.

Schilly was impressed by Jim's presentations to the community, and glad that Jim had a well-respected St. Louis construction company, Alberici Constructors Inc., involved. "I think he did an outstanding thing by hooking up with Alberici," he says, "but there was still a lot of backlash." Schilly tried to calm protestors' fears; iron smelting was not the environmental disaster that lead smelting had been. Iron ore is an oxide, and releases oxygen during processing. Lead ore releases sulfur. Lead is toxic to living things; iron is not. State environmental regulators even weighed in on the issue at city forums, stating that any such facility would meet or exceed all state and EPA environmental regulations.

Many residents completely supported the smelter and couldn't understand what the fuss was about. "The project is also popular at the Crystal Tavern, a dimly lit bar down the street from Mr. Schilly's office," reported the *Wall Street Journal* in September 2008.[3] "Owner Peggy Brown says people remember a 7 A.M. rush at the bar as nightshift employees stopped in for drinks after work. 'Everyone I talk to is for it,' she says. 'They don't understand what the big deal is; they know we need jobs and we need industry.'" The article also quoted a member of Concerned Citizens of Crystal City, Roger Leonhardt; "He views the project as a potential disaster and worked with a local opposition group, filming town-council meetings and posting relevant snippets on the Internet. He admits the town needs revitalization. 'But we'd like to shift to progressive jobs,' he says. When pressed to explain, he says simply: 'Something different than poisoning people. The problem around here is that many people think that if it isn't manufacturing, it isn't anything.'"

Concerned Citizens of Crystal City filed a lawsuit claiming that the secret meetings between Jim Kennedy and the city council violated the 1973 Missouri Sunshine Law, which requires that certain government proceedings must be held in public. However, the law actually allows closed meetings in the case of "leasing, purchase or sale of real estate by a public governmental body where public knowledge of the transaction might adversely affect the legal consideration therefor." When the plaintiffs refused to turn over the names and e-mail addresses of the anonymous posters on their website, the judge tossed the case.[4] Schilly still feels the sting of how some in his city twisted what he believed was a wonderful opportunity. "I feel to this day, and I know in my heart, we did everything above the table." Still, the drama was tearing the town apart.

The next presentation to the city on March 2, 2009, was given by Alberici's project director, Bob Niemeier, rather than Jim Kennedy.[5] To a packed crowd at the elementary school, Niemeier talked about the pipeline from the Pea Ridge mine. He detailed the billion-dollar shipping barge and the iron processing plant. Niemeier showed samples of magnetite and pig iron to the audience and described the smelting process.

Neimeier said that six hundred production workers would be needed at the site, and he expected to be up and running in four years. Many in the audience seemed pleased. A nineteen-year-old asked the council to give him and his friends something to return to after college. One woman said she wished the plant would be up and running in days instead of years. Another Crystal City resident said the proposed facility would be a boon for schools.

Still, the situation in Crystal City deteriorated. Vitriol against the project raged online and in occasional one-on-ones in the street between supporters and the opposition. Sources close to Jim say he was beside himself, putting out one fire after another. The Chinese investors from Minmetals Inc.—who were about to put in funds for the entire Pea Ridge development project—became understandably nervous. They asked to come for a site visit. To smooth things over, Mayor Schilly and members of the city council rolled out the red carpet, and the two Chinese representatives appeared to be solidly in favor of moving forward. Unfortunately for Jim and Schilly, there was a city council meeting that evening, and the Chinese investors wanted to attend. The Minmetals representatives sat in the front row of the hundred-seat room. Schilly cringed as people packed in, wearing T-shirts calling out the smelter as a death sentence. Protesters stood against the walls, muttering, and some made comments—very loudly—deriding the Chinese investors.

Just like that, Jim's financing was gone.

What transpired next can be sutured together by court documents, news stories, and a few who were caught up in the ensuing drama and legally able to speak about it. Stan Short, Jim's financing partner in these tumultuous times, offers this prelude: "This whole story—when you know as much as I do—is feature-film worthy." Short still reels from his accidental role in a saga that involved investors from around the world, dozens of law firms, and a hundred-year-old St. Louis construction company. It was a brutal three-year battle for control of Pea Ridge, the Crystal City development, and all of Upland Wing's holdings—an unceasing attack that that Stan Short says was no less than "War—with a capital W."

16

War

What brought Jim Kennedy and Stan Short together at one of the most desperate points in Jim's life was a microscopic life form for which both men had big ideas. Their meeting was set up by an esteemed professor at the Missouri University of Science and Technology in Rolla. Both men came to Dr. David Summers separately because the professor of mining engineering was experimenting on how to grow algae underground. Jim's interest in the slimy green plant cells came out of his desire to achieve something approaching energy "recycling" at Pea Ridge. When iron ore is formed into pig iron, the process releases carbon dioxide. Jim's plan was to capture the CO_2 to feed his algae, then convert the algae into bio-diesel. The ultra-clean bio-diesel could fuel the underground fleet of mining equipment. He wouldn't just save money, he'd improve the underground air quality for his workers. Short came to algae because he'd spent much of his life finding fascinating new things with the potential to make big money—and then finding people with money to make those things happen.

Before returning to Jerseyville, Illinois, to be near his daughter and grandkids, Short had a long and varied career in the financial industry. He managed portfolios for the St. Louis Union Trust Company in 1969 and then ended up in New York City in 1974, "when Wall Street was a small place with no cell phones or Internet," he says. Short developed an exceptional roster of clients—"a gold-plated Rolodex in a way." Short was soon calling on some of the biggest corporations and investors, including Texas's Bass Family and Warren Buffett. After that he worked with Peter Cohen (later Shearson Lehman Brothers CEO) on a securities start-up funded by billionaire banker Edmond Safra. When Safra wanted out of securities and dissolved the fund, Short went to work in venture capital. It suited him well because he was endlessly interested in all kinds of topics. He found financing for HIV treatments. He put together money for an Internet start-up in 1989. Later in life, he became interested in biofuels, working on projects in Turkey, Germany, and the U.S. Midwest—which was what put him across the table from Jim Kennedy in the spring of 2009.

For years, researchers had tossed around the idea of using artificial light to grow algae in abandoned mines. The earth's temperature belowground is about 50 to 60 degrees Fahrenheit once you hit 20 to 150 feet below the surface, and increases 1 degree Fahrenheit every 50 feet. Since algae grow best between 68 and 70 degrees, it's much easier to keep algae in constant bloom underground. It also helps keep out contaminants that grow in surface water. When Short met Jim for the first time, he knew only two things about Jim Kennedy: one, that he was interested in algae, and two, that he was developing the old glass factory in Crystal City for his mine in Sullivan, Missouri. What the old Pittsburgh Plate Glass Company left behind that interested Short were dozens of large caves carved into sandstone rock formations from years of mining silica. Short wanted those caves.

In a St. Louis restaurant, Short handed Jim a twelve-page printout of his algae presentation. Jim took the three-ring binder and looked at each of the slides. "He was nodding his head on every page and every bullet point," says Short. Jim knew exactly what Short was talking about. In fact, Jim had a deep interest in biofuels, and as with every topic he pursued, he'd cataloged an inordinate amount of research. To grow algae at Pea Ridge and turn it into fuel, "I took every other successful idea in the public domain," he says. The problem, Jim found, was one of volume. Because the algae starves for light six inches from a light source, growing pools take up vast two-dimensional space for what they are worth. So Jim designed vertical growing tanks that used fiber optics to light up the vertical and horizontal space. With a farm of forty-foot tanks, he could produce between 700 and 1,000 times the algae on the same amount of real estate that one "impoundment," or pond, would use. He took his idea to the university in Rolla, where he knew Professor Summers was working on underground algae. Summers told Jim that it was an excellent idea—Jim should patent it. In 2008 Jim filed a patent on his Bio Breeder System for Biomass Production.[1] But by 2009, because of the enormous amount of work it had taken just to get Pea Ridge engineering and financing in place, biofuel was on the back burner.

The two bonded quickly, as people who share an obscure obsession often do. Jim opened up to Short. "Now let me tell you about my project," he said, and he began to share the story of Pea Ridge. Jim told Short of his plans to bring the mine back to life after the price of iron ore exploded in 2006. "The project was amazing and huge," says Short. And very costly too. Jim told Short of his money woes with Minmetals dropping out, and how after the economy crashed in 2008, he hadn't had any luck finding new investors.

By the time Jim finished his story, Short was completely engaged. "I realized that in order for Jim to do anything with respect to the river properties, he had to have some new money and get his project back in gear," says Short. And if Jim didn't have the Crystal City properties, Stan would not have access to the sandstone caves. "So I made the determination on the spot. I said, 'Look, this sounds very interesting. I'd be happy to help you do whatever I can.'" Jim immediately agreed. On April 8, 2009, they signed a contract. For starters, Short would find $20 million to help pay off Upland Wing's debts and jump-start the engineering studies. Then he'd help gather the rest of the funding for the project. Short's take was 5 percent.

For the next year, Short worked full-time on Pea Ridge financing. He had an extensive list of possible sources to target, and seven consultants to help him make connections. To finance something as big and complicated as the Pea Ridge project was very challenging. There were meetings and reports and analyses. Short reached out to dozens of U.S. prospects in New York, Houston, Connecticut, St. Louis, California, and New Jersey. He queried international investment firms and large commodities companies in the UK, Singapore, India, Germany, and Canada. He contacted gargantuan players like Bank of America and niche investors like Resource Land Holdings in Colorado. It was exhausting, and he loved it. Within a year, Short had logged thousands of hours, and by the spring of 2010 he had some significant interest from a few major investors.

In the fall of 2010, with the help of a consultant, Short landed a beautiful deal. Glencore—at the time, a $60 billion company and the largest commodities trader in the world—wanted to fully develop Pea Ridge.[2] On

October 18, Glencore and Wings released a joint press release. "The partnership between Wings and Glencore is still in its earliest stages and is dependent on securing 'financing on competitive terms' and the completion of commercial and technical feasibility studies. The studies will begin in the next 60 days and could take up to eight months to complete," the release announced.[3]

Three weeks later, on November 5, 2010, Alberici, the company that Jim had hired to work on the pipeline project, filed a motion for a preliminary injunction to stop the deal.[4] If Glencore was allowed to partner with Jim and Nina, that would leave Alberici out of the equity they wanted in the mine. Alberici's role would be "at a minimum, diluted or conceivably become worthless," the injunction argued. On February 22, 2011, a county judge upheld the injunction. There could be no deal between Wings and Glencore.[5]

Apparently, the parties had been sworn to secrecy, because Stan Short was not privy to any of it. The Glencore deal that he'd worked on for so long disappeared without a trace, leaving him only able to speculate why. For years he thought it was because of a giant merger that Glencore had in the works, or perhaps because the company was about to go public on the London Stock Exchange. (Like Jim and Nina, the Glencore representative who worked on the Pea Ridge partnership will not discuss what killed the deal.)

Still not overly concerned, Short went to find another funding source that would pay off Alberici for work they'd done on the mine and finally get the mining project back on track. He was delighted when he found a financier who would provide $20 million to cover outstanding liabilities, complete facility upgrades, finish the feasibility engineering, and then fund the entire project once the engineering was complete. But just as with Glencore, the deal evaporated—mostly in secret—in a small Washington County courtroom. According to Stan Short, the following year was tumultuous. When all was said and done, the mine that Jim Kennedy loved so much no longer belonged to him. In January 2012, the judge ruled that Pea Ridge and all of Wings' holdings now belonged to Alberici.

At a town hall meeting in Crystal City three weeks after the judgment, the *St. Louis Post-Dispatch* reported that Jim and Nina sat quietly and listened to Alberici project director Bob Niemeier present a slide show about his company's plans for Pea Ridge and Crystal City. Niemeier described how Alberici would make pig iron in the smelter, the port they would build to ship to steel manufacturers along the river, and the underground pipeline to carry the slurry from Pea Ridge. When a reporter asked what the difference was from Jim Kennedy's plan, Niemeier said, "We're not looking for investors,"[6] and that Kennedy and Abboud would no longer have any involvement in or control over the operation of the mine or the property leased from Crystal City. Also different this time around: there were no protesters. The meeting room was quiet and only partially full. Jim would not answer questions from a reporter for the *St. Louis Post-Dispatch*, except to say that he came to the meeting to show his support for the project. "The city needs this," he said.

Mayor Tom Schilly was there too. He says he has no idea what happened between Jim and Alberici, but he was sorry to see Jim off the project. "I stuck by my guns, and I stuck all the way through," he says. "There were times when I wasn't able to speak to Jim because of the lawsuit that was going on, but I've got a lot of respect for him. I just really wish it would have worked out." Schilly says that he never envisioned that anything would happen on the old property in his lifetime, "but Jim had it all mapped and got the permits and the covenant restrictions out and everything. I don't want to pretend I know what happened between him and Alberici, because I don't. But Jim did spearhead it. He had a vision, that's for sure."

For Ben Auld, the mining engineer from Australia who greatly looked forward to the Pea Ridge project, it was a very sad day when Jim Kennedy lost the mine. "Clearly it was hard for him to take, but at the same time, he told me he believed he was right. He told me he still believed in the project." In the meantime, Jim told Auld he would focus his attention on becoming a rare earth expert and use his energy in other areas.

Many of those involved in the project were shocked when somehow, after the trial, with all the cards stacked against him, Jim Kennedy walked away with a $13 million "sale" of the mine to Alberici.[7] Jim and Nina would also earn a 2.6 percent royalty from the Pea Ridge iron ore—a percentage that could bring in millions of dollars a year once the mine went into production.[8] Even more surprisingly, Jim and Nina took away 70 percent of the rare earth elements and all the nonferrous minerals. On January 11, the *Jefferson County Leader* interviewed Jim: "I am no longer involved in the iron ore part of the business," he said. "I can't comment beyond that. I do retain some interest in the PPG property, but I intend to pursue the rare earths elements and nonferrous metals side of the business."[9]

Stan Short read of Jim's coup a few months later. "Everybody who was involved was curious what the agreement was, what else it contained, and how it came about," he says. Once again, Jim Kennedy had pulled a rabbit out of his hat. Where that rabbit came from, he won't say.

In the spring of 2012, the economically depressed towns of Sullivan and Crystal City, Missouri, waited anxiously for their twenty-first-century renaissance. Alberici was one of St. Louis's most renowned corporations. Their reputation was stellar. They had the expertise and the money to make it happen. Washington County, the home of Pea Ridge, and Jefferson County, where the old Pittsburgh Glass Company sat abandoned, would share over a thousand new jobs. Former Pea Ridge engineer Larry Tucker and other old-timers signed on with Alberici's start-up crew, grateful that their mine would prosper again as it had in the days of St. Joe's. "Jim Kennedy had the vision," says Tucker, "but Alberici had the money."

In January 2012 Alberici partnered with a Canadian company called MFC Industrial, LTD to market the iron from Pea Ridge. In April 2013, MFC's CEO announced that the project was "full steam ahead."[10]

They never broke ground.

By 2014, Alberici's name adorned the old St. Joe's offices in bold red letters, but the only thing in operation was the processing plant constructed by Jim Kennedy's scrappy team in 2009.[11] Only weeds grew on the Pittsburgh Glass property in Crystal City. "Nobody's moving on it," says Mayor Tom Schilly, who—in his mid-sixties—misses his vibrant American Midwest. Even after everything that happened, he remains hopeful. "We're sitting on 249 acres of prime industrial real estate. We still have the property, and Alberici still has the lease. We'll see what happens, and we'll go from there."

In August 2014, MFC boss Michael Smith told investors, "In this country, many things tie our hands. There are many regulatory requirements."[12] However, American regulatory requirements were not the reason that the Pea Ridge project remained at a standstill. The three-act tragedy that played out between 2006 and 2014 in the mineral-rich mountains of southeastern Missouri was 100 percent orchestrated by the People's Republic of China. Just as China's building boom reverberated across the globe in 2006, reviving mothballed mines from every continent, so did the slowdown of its building sector. In the spring of 2015, iron ore fell to $51 a ton, down from its height of $187 in 2011.[13] The futures market looked even more dismal. And in Sullivan, Missouri, the fight over Pea Ridge between Jim Kennedy and Alberici Constructors became just another cautionary tale in the history of Missouri mining.

17

A Meeting of Minds

There was one thing that kept Jim Kennedy sane during the tumultuous years that stripped him of his beloved Pea Ridge. In the midst of a near nervous breakdown fighting Alberici and their army of lawyers, Jim found relief and a larger sense of purpose. As bad as it was to lose something so dear, a gigantic storm was brewing that dwarfed his few square miles of heartache in the Missouri mountains. The very foundation of the United States of America was crumbling, he believed. And somehow, every fiber in his being was telling him that he had a responsibility to help fix it.

One unusually warm October morning in 2009, Jim arrived at Reagan National Airport in Washington, D.C. He was running late and flustered because he'd missed his first flight from St. Louis while sitting right at the gate, tweaking his PowerPoint presentation. "I looked up, and everyone was gone," he remembers. As usual, he called Nina, who found the next flight and sent him to the right gate. Jim had met with dozens of scientists over the last few years. But this meeting would be different. The cab ride gave him a few minutes to reflect. Here he was, a guy who'd slept through most of high school—who'd never taken a physics or chemistry class—on his way to pitch a monumental idea to the Pentagon's strategic materials expert, Rick Lowden. Lowden was on loan from Oak Ridge National Lab to help the Pentagon develop national strategies to deal with critical materials.

In order to hook Lowden into his plan, Jim explained that over the past decade, a mixture of globalization and bad political decision-making had put the entire U.S. defense industry and the rest of the Western world as well as Japan at the whim of the Chinese. On his laptop, Jim showed photos of cruise missiles, smart bombs, Predator drones, lasers, launch vehicles, night vision, and everything else utilized in warfare, all created by Fortune 500 American defense contractors: Raytheon, General Dynamics, Boeing, Northrop Grumman. Above each photo, Jim listed the rare earth elements that were essential to manufacturing that defense product.

He told Lowden that the Chinese had built a monopoly around rare earth elements. There wasn't a single way the United States or any other country could access the metals needed for several trillion dollars' worth of end products without going through China. China could shut down exports on a whim and ultimately cripple the U.S. Armed Forces. While Jim made his case by concentrating on the defense industry being the most dangerous issue, he believed the problem was much bigger. "More important even than defense products," he says, "our economy can't survive without being plugged in to the technology and manufacturing economy." The United States, Germany, Japan, France, Israel, Sweden, Korea, and Switzerland, as well as every other country in the world that produces consumer products with rare earth elements, are now at the mercy of the Chinese. Without the ability to obtain and use rare earths, the United States could never revitalize its manufacturing economy. "And you can't have a military," Jim told Lowden, "if you don't have a tax base to support it."

But Jim didn't just come to state the facts. He came with an intricate plan he'd refined over several hundred research hours. He modeled his strategy after the way the U.S. government handled the semiconductor shortage in the late 1980s, when Japan was dominating the industry. American chip corporations had convinced Congress that being at the top of the semiconductor game was a national security issue, so they pulled together a consortium of fourteen chip manufacturers, including Intel, Texas Instruments, along with the Department of Defense, and the Defense Advanced Research Projects Agency (DARPA), to form what they called Sematech (from Semiconductor Manufacturing Technology). They went into operation in 1988. The DoD granted Sematech $100 million a year, and U.S. chipmakers regained the lead in the 1990s.[1] "It was the same model that ended the bankruptcy and starvation of eighteenth-century American farmers and turned the USA into the most productive and efficient producers of food in the world," says Jim, "an old-fashioned cooperative—real old-school stuff."

To Jim, the model made perfect sense. The Pentagon and Congress had recognized that a loss of computer chip technology was a national security issue; surely they would buy into his idea as well. "My argument was for a cooperative to handle the mining, processing, and production of rare earths into value-added products." The solution also fit with current political realities of the time, with lawmakers vehemently opposed to any new spending.

Jim knew the climate. And for that reason he designed his theoretical cooperative to be funded, owned, and operated by the end users of rare earth materials. The partners in the co-op, Jim told Lowden, would include giant corporations like Siemens (Germany), Boeing (United States), Toyota (Japan), Apple (United States), Thales (France), and BAE Systems, the UK's largest defense contractor. All were currently completely dependent on China for their rare earth metals, alloys, magnets, and components. In Jim's scenario, the cooperative would effectively break the Chinese monopoly.

At the end of the meeting, Lowden told him that he thought his plan made "a lot of sense."

Jim was thrilled. "What other things are you looking at to solve this national security problem?" he asked. Lowden mentioned a mothballed California rare earth mine called Mountain Pass. The mine's owners—a company called Molycorp Inc.—were talking about getting back online. Molycorp was lobbying hard in Washington for guaranteed government loans to restart their mine.

Jim had already done his homework. Hearing that Molycorp had an interest in Pea Ridge, Jim and Nina had gone to the desert in 2007 to check out the Mountain Pass facilities. "We wanted to see if there was any potential for working with the Molycorp folks," says Jim. "It was the first time I had ever been to an open pit mine," Nina remembers. "What really stood out was the feeling of it being an abandoned facility. There were maybe less than a dozen employees milling around, no real activity that I was aware of." To Jim, the operation was quite unimpressive. "Inside, there was an old processing facility and rusted and corroded equipment. Other than some polyethylene tanks, none of it could be reused." Jim asked for a sample of bastnaesite ore as a souvenir for his geology mentor, Cheryl

Seeger. When he got home, he dug deep into a half-century's worth of scientific studies and commercial operations from Mountain Pass. "What I discovered was that the mine was sadly deficient in many ways."

Jim knew that Mountain Pass could do nothing to solve the rare earth crisis for the Pentagon or anyone else, and he told Lowden as much. "I told him that Molycorp's geologic deposit lacked most of the critical rare earth elements necessary to fill our national security needs. This glaring fact angered me. Everyone promoting Molycorp ignored it, failed to understand it, or blatantly covered it up. It was difficult to believe that someone like Rick Lowden did not understand it." Jim asked Lowden if he had some other information, and Lowden assured him that he didn't. "I told him there was no way that Molycorp could compete or even survive against China's monopoly."

"But what else do you have on the table besides Molycorp?" Jim asked. "I'm sure the DoD has its own game plan to assure these resources are coming in." Lowden looked at him like he was nuts, and Jim asked again. "There's got to be some big secret Pentagon plan, or someone in the private sector who is working on this?" Lowden told him that no, Jim's plan was the only one he'd seen. Jim was incensed. His feigned inside-the-beltway cool instantly fell away. "You have got to be kidding. You can't run a department of defense like that!" Jim was almost yelling. "'This is a national security issue,' I told him. 'This isn't how things are supposed to happen. You can't just count on a solution showing up on its own.' And he says, 'Well, *you're* here, aren't you?'" Jim stared at Lowden, speechless.

Jim left the meeting with a heavy heart. "Honest to god, there was no alternative plan. Nothing. I was just shocked. That's the first time I realized we are a rudderless ship, completely adrift in the ocean."

Jim had one more item on his agenda on this trip to D.C. He'd agreed to meet with a group he knew nothing about, except that they called themselves the Thorium Energy Alliance. The only reason he agreed to the meeting was that thorium was a serious problem for the rare earth industry, and the missing piece in the puzzle that he'd been racking his brain to solve.

The U.S. Nuclear Regulatory Commission (NRC) and the International Atomic Energy Agency classify radioactive "source material" as any material containing more than 0.05 percent uranium or thorium.[2] Nearly all high-value heavy rare earth minerals contain thorium and/or uranium. Monazite has thorium levels that are literally hundreds of times above the threshold. Regulations weren't strictly enforced before the 1980s, when the international community signed on to NRC regulations. Before then, geologic deposits called heavy mineral sands were the original source of rare earths. The sands come from the weathering, over millions of years, of igneous and metamorphic mountain ranges. Eventually, the resulting sand and clay particles become beaches and coastlines rich in heavy minerals. The deposits are mined mainly for valuable titanium and zircon.[3] In the United States, heavy mineral sands are mined for titanium in Virginia and Florida.

For mine owners in the United States who have rare earths they want to market as byproducts of their primary ore, getting licensed to deal with "source material" is onerously expensive, and represents an unlimited liability. If a mine processes rare earths and piles up the thorium on the side, this is considered long-term storage of unutilized "source material," which then is automatically redefined as nuclear waste. The nightmarish costs and complications around the disposal of nuclear waste stymie the U.S. Congress and the Department of Energy. Yucca Mountain, a proposed waste depository gouged out of a Nevada mountain, was finished in 2002 but remains empty to this day. The cobwebbed caverns are a monument to the political and economic turmoil associated with source material and nuclear waste. No private company wants to get caught up in it, or even think about picking a fight with the Nuclear Regulatory Commission in a country afraid of even the slightest natural radiation. Rather than taking on the "source material" liability, mines quietly bury valuable rare earth minerals or send them to tailings lakes. Jim found this an idiotic policy, because the threshold for allowable radioactive material was ridiculously low.[4] "You can lounge on a public beach in Brazil that has thorium levels well above ten to 200 times the threshold." Pea Ridge was rich in heavy rare earths, but plagued with thorium as well—hence, Jim's thorium problem.

On October 19, after meetings with congressional aides to get backers for his rare earth cooperative, Jim arrived in the conference room at Gallaudet University. Fifteen years in the financial business and elite business school could not have prepared him for John Kutsch's collection of thorium nerds. The brainpower was so intense that the air felt heavy with cerebral momentum—as if everyone was constantly doing theoretical calculations in their enormous cortices. There were scientists from Los Alamos, Idaho National Laboratory, and several universities. There were nuclear engineers from Canada and around the world—even a former member of the Japanese parliament. They all had one thing in common. Like John, they were passionate about reviving thorium nuclear energy research and building safe and clean nuclear power plants to help reduce the impact of greenhouse gases. Jim wondered what the hell he had gotten himself into. He had no idea what he could possibly bring to the table.

John's introduction was decidedly awkward, since he still didn't know what the dude wearing jeans and cowboy boots would bring to the table either. Jim got up to speak. "I explained how deep and broad the rare earth failure is for the United States and its allies," Jim says. As with Rick Lowden at the Department of Defense, Jim skewed the talk to rare earth defense applications, "because it's the only thing that gets people to wake up," he says. "So I'm kind of pounding on the military stuff and how hard it would be to fix it, and how this problem is all chemically and mineralogically linked to thorium. When I finished, everyone's hair's blown back. They're all physicists and engineers, and most of them have no idea about mining."

"It was true," says John Kutsch. "I'm ashamed to say that as much as I knew about thorium and the molten salt reactor by then, I had no idea where you got thorium from."

"So at your mine, you're producing rare earths?" one scientist asked.

Jim told him, no, historically the mine could not produce rare earths because of this thorium problem. "But the problem is much larger than the mining project I control," he said. "It is this thorium problem that killed off the entire Western supply of high-value heavy rare earths. All of the world's suppliers just stopped feeding the U.S. value chain. Eventually the U.S. value chain went bust and relocated to China."

"What if you could fix the thorium problem?" one person asked. "How much thorium byproduct could you bring back into production?"

In his mind, Jim did a little calculation. "About seven thousand tons per year of thorium." To the audience, the amount was staggering. "These guys are absolutely flabbergasted," says Jim. "They said, 'Do you realize that's enough thorium to generate the world's electricity needs for a year?'"

"Part of my brain is telling me these guys are all nuts," Jim says. "Another part is saying, okay, some of these guys are from Los Alamos, Idaho National Lab, and the Canadian and Japanese nuclear industry . . . they must know what they're talking about."

Jim sat back and tried to comprehend the next talk. The presenter discussed what sounded like some fantasy nuclear reactor from *The Jetsons*. Everyone in the audience nodded and affirmed the speaker's every point. Jim was completely out of his element. Were they all wearing invisible aluminum hats? The conference break was even more intense. Jim was clueless at how to converse with these guys (there are very few women in nuclear engineering in Western countries, and none attended John's first conference). Again, Jim relied on his military training. These scientists and engineers were clearly on a higher cerebral level than he'd ever experienced. "If I wanted to adapt, I had a lot of work to do," he says. "I collected business cards and made a commitment to myself to investigate what they were all so excited about." Back home, he read everything he could on this "fantasy" reactor. "I discovered that the U.S. government had built and operated the reactor successfully in the sixties. It was real."

While Jim digested what he'd seen at the conference, John went home and tried to wrap his head around the issue of rare earths and China. "I knew exactly what rare earths were. I'd spec'd out all sorts of rare earth stuff. If we needed a very high-energy motor to drive something, we would specify a rare earth motor. If we needed a laser for doing some distance calculation, we would specify a YAG [yttrium aluminum garnet] laser."

"You don't have to be terribly technical to wrap your melon around what a nightmare the Chinese monopoly is," John says. "What's appalling is that our fearless leaders can't wrap their heads around the idea that we'll be screwed if we don't control our supply chain of raw materials." Jim Kennedy was John's greatest priority to follow up after the conference.

A few weeks later, John invited Jim to a meeting at the University of Illinois at Urbana-Champaign. Several thorium colleagues, including the head of the university's nuclear program, were brainstorming about how to get Gen 4 (meaning fourth-generation reactors, with a design completely different from standard light-water "Gen 3" reactors) nuclear programs some much-needed attention in Washington. Afterward, the group hit Murphy's Pub, a local watering hole.

Jim was very eager to speak with John. "I needed to know his plan," he says. "I'd formulated one myself, but if John already had a plan, I wasn't sure what my next move would be." The waitress brought a pitcher of tasteless draft beer, and Jim took the opportunity to invite John to the bar for a Sam Adams.

Jim was a bit nervous when he finally had John alone. "I told him that I was fully behind Thorium Energy Alliance goals, but Congress, the DoE, the Nuclear Regulatory Commission (NRC), and the entire energy industry would block this—no, kill it."

"How are you going to get thorium energy developed in a world of opposition?" Jim asked.

It was a question John hadn't a clue how to answer. "I had no idea how to bridge the giant valley between where we were and where we needed to be," he says. So if someone had a better idea, he was certainly willing to listen. He nodded for Jim to continue.

"John, it will never happen without some sort of cover." Jim excitedly laid out his plan. Rare earths and thorium are connected at the mineralogical and geopolitical level. No member of Congress would champion new nuclear energy legislation, but nearly every member would support a solution to the rare earth crisis—a demonstrable national security issue.

John was impressed by Jim's creative way of putting the pieces together. He also bonded with Jim because the two of them wore their patriotism to America on their sleeves. Both were determined that America should lead the world in technology and manufacturing again. Neither man spoke of making money off their ideas, or forming a for-profit corporation—only of how they would fix a broken system with stealthy brilliance. "It was an excellently thought-out plan and withstood years of tweaks to remain essentially the same today as then," said John in 2015. As close as they would become, it would have been apropos if they'd cut their wrists and bonded in blood right there at Murphy's Bar. Instead, Jim says, "We shook on it, then walked back to the table to join the others."

And that was it. From that day forward, Jim Kennedy and John Kutsch were closer than Siamese twins: one fighting to bring back defense and commercial manufacturing to the West and its allies, one determined to see his country at the forefront of a nuclear renaissance—both completely dedicated to each other's goals. "With John's help, I could actually build a strategy that would make sense," says Jim. "Before, I could talk all day about the rare earth problem, but until you fix the thorium problem, you're not going anywhere. John had the other half of the equation. We could make this happen."

Together, they went out into the world, armed with facts and charts and numbers and their plan to save the country. They studied history, politics, science, and current events and talked endlessly about ways to make what they knew was possible a reality. Over the next several years, the two of them would travel tens of thousands of miles, talk to countless politicians, bureaucrats, scientists, and other advocates. Both their bank accounts slid deeply into the red, and both their wives half-seriously threatened divorce. But what kept them on their quest above all were their very similar hearts.

"For Jim and I and others to try to turn this ship around may be a fool's errand, but I really don't have much choice," John says. "Once you know there is a better way, it is a moral decision to pursue it." He calls himself "cursed" to be brought up by the most honest man in the world—his father. "Men for others"; "To those to whom much has been given, much is expected": John gave these credos, learned in Jesuit school, great deference. "So my lot in life is to try and build in the face of those who would dismantle."

18

Rough Waters

A year after Jim Kennedy's meeting with the Pentagon's point person on strategic materials, Washington finally woke up to the reality of China's rare earth monopoly. The story began one September morning in 2010 aboard a rusty fishing trawler in the East China Sea. Zhan Qixiong, a forty-one-year-old Chinese skipper, captained the *Minjinyu* near a rocky volcanic island chain, much as his ancestors had for centuries.[1] The seas were choppy, but the weather was fair, and the experienced seamen cast a wide net behind the vessel, where they hoped to catch mackerel, croaker, herring, and other fish plentiful in that warm latitude. Seagulls followed the boat, looking to score an easy meal, as the crew slowly hauled in the net. In a video recording from that day, you see fish desperately trying to fling themselves back into the sea. A closer look into the dragnet of that rusty Chinese fishing boat reveals a brewing political storm involving the world's technological superpowers.

Back in the 1850s, Japan claimed the Senkaku Islands, but both the Chinese and Taiwanese—who called the islands Diaoyudao (the Diaoyu Islands)—argued that this strategic territory had been theirs since ancient times. The islands came briefly under U.S. control after 1945. Then in 1951, Japan and the United States signed a treaty turning over jurisdiction of the islands to Japan. China declared the treaty illegal, but Japan's advantage solidified in 1971 when the United States handed over Okinawa and drew the Senkaku Islands into the deal.[2]

It was this twentieth-century political maneuvering that gave the Japanese coast guard jurisdiction to patrol the East China Sea that September morning. Zhan Qixiong—who believed that he was fishing in Chinese waters—must have felt some irritation when he saw the giant

Japanese coast guard frigate approach from the distance. One of the *Minjinyu* crew looked out toward the patrol boat with a pair of binoculars as the frigate blew a loud horn, its baritone sound waves attenuating across the water. Looking mostly unconcerned about the Japanese patrol boat, the crew of the *Minjinyu* continued their work, with several crewmembers carefully pulling up the net and others gathering the catch. The crew likely felt little threat. There had been lots of these cat-and-mouse games since 1971—big white patrol boats chasing down aging Chinese trawlers. Chinese captains mostly ignored the patrol boats, and then powered slowly away. To the Japanese crew, this day probably seemed as if it would end no differently. As the fishermen pulled in the net, the frigate approached *Minjinyu*'s starboard side. One of the coast guard crew yelled in Chinese over a loudspeaker, and a few other Japanese crewmen shot video of the scene. In it, after a while, the fishermen lazily pull the net into the boat. It looks as if the fishing boat is leaving, but instead, Zhan Qixiong motors the trawler in a slow semicircle and rams into the bow of the coast guard ship.[3] It was a significant bump that shook both steel hulls. On the video, the Japanese can be heard yelling—clearly shocked—as the fishing boat powers away.

The next day, the Japanese coast guard went after the *Minjinyu* again, this time with its siren screeching into the wind, cutting Zhan off with a sharp turn into the trawler's bow and causing what appears to be an unavoidable collision with the coast guard ship.[4] The video shows clear damage to the patrol boat as black smoke pours from its starboard hull. The coast guard members boarded the *Minjinyu* and arrested the crew. Still in grimy shorts and wearing flip-flops, Zhan and his crew were taken and detained at Ishigaki Island.[5] The following day, the thirteen crew members were released, but Japanese authorities held Zhan for obstruction of performance of public duty and illegal fishing—charges that carried up to three years in prison and a 500,000-yen fine (U.S. $4,000).

The fallout of the arrest reached a grand scale. Chinese fishermen staged protests in the street, burning Japanese flags. Chinese leaders canceled meetings with Japanese ministers. Animosity between the prickly adversaries grew. China's premier, Wen Jiabao, demanded Zhan Qixiong's release "immediately and unconditionally."[6] Japanese authorities refused for two weeks, and then finally backed down. What was widely reported next was a highly aggressive move by China in retaliation for the incident. It would be what analysts called a calculated and hostile strike against the economic interests of Japan. And the story, repeated over and over, would send shock waves through Tokyo, Washington, and the world.

Keith Bradsher, *New York Times* bureau chief in Hong Kong, wrote, "The Chinese government has blocked exports to Japan of a crucial category of minerals used in products like hybrid cars, wind turbines and guided missiles." Bradsher quoted "industry officials" as his sources.[7] The Agence France-Presse news service quoted a spokesman from a Japanese trading house who claimed shipments of rare earth metals from China would be cut on September 21. Dudley Kingsnorth, an influential consultant for rare earth interests outside China, told the *Times*, "By stopping the shipments, they're disrupting commercial contracts, which is regrettable and will only emphasize the need for geographic diversity of supply."[8] Economist Paul Krugman followed up in the *New York Times* with a harsh condemnation of China's actions. "The incident shows a Chinese government that is dangerously trigger-happy, willing to wage economic warfare on the slightest provocation . . . What you have is a portrait of a rogue economic superpower, unwilling to play by the rules."[9]

China rejected the reports wholesale. And in fact, post-hype, the Harvard-MIT journal *International Security* analyzed imports reported into Japan from China during that period and found no evidence of any embargo at all.[10] But the China-baiting, done purposefully or not, by Western and Japanese media, or by industry consultants tied to mining interests, was a highly productive mechanism to get the rare earth industry some much needed publicity. China would benefit from the false narrative; manufacturers in Japan and the United States, fearful they wouldn't be able to get rare earth components, had even more reason to send their operations to China. Washington gasped. How could China do this? The nation's offensive move proved to the world that China could completely shut down enormous sectors of a nation's economy at will. Beijing continued to deny that there ever was an embargo, but their protests were mostly ignored as leaders in Washington and Tokyo—who'd previously turned a blind eye to China's systematic takeover of the entire rare earth industry—were now wide awake and in shock at China's hostility. China joined the World Trade Organization, didn't they? This was totally uncalled-for.

While the East China Sea collision and bogus embargo brought some much-needed hype, the real story that turned the tide on rare earths was more complex and calculated. A year earlier, in 2009, China roused the ire of the tech manufacturing industry when Chinese officials decided they would cap exports of rare earths "out of concern for the environment and to minimize pollution," Liu Aisheng, director of the Chinese Society of Rare Earths, told Bloomberg News. While processing of rare earths is quite toxic because of the harsh chemicals used to decompose the minerals, few believed this was China's main reason for cutting exports.

Prices for neodymium, dysprosium, and samarium (all used in magnets and batteries), ytterbium (used in lasers), and other rare earths with assorted technical applications shot through the roof as China cut exports by 70 percent. Tech companies still making products in their own countries suddenly realized that there wasn't a single mine in the world outside China extracting and processing rare earths. It was exactly what Jim Kennedy had warned Rick Lowden about at the Department of Defense the year before. For manufacturers, the price of rare earths wasn't the problem—only infinitesimal amounts are used in end products. What was highly worrisome was getting access to them at all. Business leaders ran to ask Washington to intervene, force China to quit restricting exports, and push the World Trade Organization to file a case against China. On September 21, 2011, John Galyen, president of Danfoss North America, an electronics manufacturing company in Baltimore, Maryland, testified before the U.S. House of Representatives Committee on Foreign Affairs, Subcommittee on Asia and the Pacific, "As a result of the reduced export quota, manufacturers like Danfoss are experiencing 800 percent increase in the price of magnets used in our compressors and protracted lead times, going from eight weeks to six months." He called for the World Trade Organization to pressure China to increase exports until other sources could be brought online, and asked that Congress and the Administration "act on this decisively to protect American industry, our national balance of trade, and our economic and technological future."[11]

In March 2011, CEO Bruce Chai of Crystal Photonics Incorporated, a manufacturer of medical imaging components, wrote to Florida senator Bill Nelson, a member of the Armed Services and Commerce Committees:

> Like other American companies, CPI relies on rare earth minerals. Matter of fact, CPI's product depends entirely on the steady supply of two rare earth minerals, lutetium and yttrium. If by whatever the reason CPI can not obtain the rare earth minerals from China, CPI will have no choice but to shut down its factory in Florida. CPI would be forced to move its facility to China to have access to rare earth minerals.
>
> This is not a hypothetical issue. The rare earth mineral crisis is an issue CPI has faced for several years and China is ready to take advantage of it. The Chinese government has approached CPI about relocating its plant to China. China does not want to be used solely for its natural resources. It wants to be a leader in high tech business and it targets companies like CPI to bring technology and jobs to China. If the United States fails to overcome this serious supply issue, CPI may be forced to take its business, technology and jobs to China. I'm sure that there are many other U.S. companies facing the same obstacles and choices that we do. It is in the hands of Congress to resolve these issues.[12]

The story of China's rare earth "embargo" and Washington's freakout couldn't have come at a more opportune time for a former Unocal executive named Mark Smith. With some of the metals trading up as much as 700 percent, in July 2010 Smith convinced Wall Street that he could capitalize on China's trade restrictions, make a bundle of cash for investors, and save the future of American defense and manufacturing industries. Smith's ace in the hole to make it all happen was a shuttered mine in the California desert called Mountain Pass. From the halls of Congress to the New York Stock Exchange, Smith traveled, swearing up and down that this old rusting mine would save the day.

The property that Smith planned to resurrect was the world's first exclusive rare earth mine, with roots in the mid-nineteenth century. As the gold rush waned along the western slope of California's Sierra Nevada in the 1860s, a horde of optimistic prospectors fanned into the Mojave Desert at the pointed tip of Nevada's southern border. There, they established dozens of silver and lead mining claims near Clark Mountain, a stratified Cambrian and Jurassic uplift forming the highest peak in the Clark Mountain Range.

In 1861, after a trapper named Johnny Moss discovered gold in the area, a second wave of prospectors followed.[13] By 1870 bountiful veins of pure silver had been carved out of Clark Mountain's north slope. Over the following twenty-five years, mines in the area produced about $5 million worth of silver (about $90 million in 2016 dollars), until it was no longer economical to take silver out of the mountain. By 1890 the miners had packed up and left to look for other deposits. A revival of the district came in the late 1890s, with construction of the Union Pacific Railroad, and an enormous mining boom followed during the war years of 1917–18. Prospecting and production reached the highest levels ever seen in the Clark mining district as copper, lead, and zinc prices skyrocketed. Prices fell after the war, and miners abandoned the area.

Three decades later, in March 1949, an engineer named Marty Hess arrived in Goodsprings, Nevada, hub of the Clark mining district.[14] The atomic age brought about a new mining boom, and Hess was promoting training seminars on the latest prospecting craze—the search for uranium. Speaking at the Goodsprings schoolhouse, a structure erected in 1913 for the children of miners, Hess piqued the interest of two in the audience: an engineer named Herbert Woodward, and motel and gas station owner "Pop" Simon. Simon offered to put up money for a Geiger counter—an instrument to detect radioactivity—in exchange for a percentage of the venture. Woodward hiked all over Goodsprings, waving the wand on his new instrument, but found nothing radioactive at all. He was about to give up when a friend with a mine in the Clark Mountains asked him to check some of his rock samples. To Woodward's delight, the lead and gold ore samples were radioactive.

Positive that he was on the verge of a big uranium find, Woodward and his wife traveled into California from Goodsprings through a deep valley cutting the Clark Mountain Range. Early prospectors named the area Mountain Pass because it was, at 4,730 feet, the area's lowest elevation and the only suitable crossing between California and Nevada in the region. Arriving at the Sulfide Queen mine, Woodward waved the Geiger counter wand over the massive tailings pile. The sensor detected more than four times the background radioactivity, sending the instrument into a fit of sizzling static.

Since the Sulfide Queen was already under claim by Woodward's friend, Woodward hiked around the area to the northwest. On April 23, 1949 (which happened to be Woodward's birthday), he found intense radioactivity about a mile from the Sulfide Queen shaft.[15] There he staked what he called the Birthday claim. Woodward collected a few radioactive rocks containing a strange brown mineral. It was not pitchblende, the uranium ore Woodward had hoped to find, but something he'd never seen before. He mailed a sample to the U.S. Bureau of Mines in Boulder City, Nevada, where a spectroscopic test identified the rare earth elements cerium, lanthanum, neodymium, and europium in an uncommon mineral called bastnaesite.

Once again, Woodward's hopes for a uranium payload were dashed. The radioactivity was coming from small amounts of thorium in the bastnaesite, not from uranium. A second analysis by the U.S. Geological Survey laboratory in Washington, D.C., found that the bastnaesite contained between 8 and 12 percent rare earths. Although there was little use for the elements themselves at the time, it was geologically fascinating to find rare earths in such high concentrations, and the USGS decided to embark on a mapping project of the area in 1949.

"It was the golden age for the U.S. Geological Survey, back when it was run by scientists and not bureaucrats," says Dan Shawe, who was assigned to make a plane table map of the Sulfide Queen mine, the Birthday deposit, and the surrounding area. "I was twenty-four. It was my first assignment with the survey, and I had no idea how to make that kind of map." Shawe, who was from a mining family in Nevada, went home and asked his wife to help him figure out the geometry of the special map projections. After completing what he thought was a good representation of the rocks, Shawe joined two seasoned USGS colleagues in the field. Over the next two years, Shawe and his colleagues traversed the high desert of the Clark Mountains, blazing hot in the summer and freezing in the wintertime. Shawe loved the work in the vast Mojave, hiking across the mountains armed with his trusted Brunton compass, using aerial photographs and topographic maps to guide him. "I'd been outdoors all my life—since I was a kid in Nevada," he says. "I really liked the time I got to be alone in those outcrops." When the fieldwork was over, Shawe felt a tinge of melancholy at leaving Mountain Pass for other projects. In 1952 his team published the world's first geologic study on a deposit of rare earth elements, *Rare Earth Mineral Deposits of the Mountain Pass District, San Bernardino County*.

Soon after, at the urging of a metallurgist who knew Shawe's work, the Molybdenum Corporation of America bought the Birthday deposit and other claims in the area. On January 16, 1951, the company put out a press release. It would soon open the world's first working mine for rare earth elements. Perhaps the metallurgist had some kind of premonition of things to come. He certainly must have had great powers of persuasion, because the deposit, although geologically interesting, was essentially a huge pile of rocks that had no real value at the time.

Mountain Pass was never a gold mine, but it did relatively well over the years. Its two main products were the only commercially used rare earths at the time: cerium, used in ultraviolet glass and lighter flints, and lanthanum, used for special camera lenses and glasses and eventually as a catalyst in the petroleum industry. The mine processed the ore into rare earth oxides (cerium oxide and lanthanum oxide) to sell to a secondary processing facility in France or China, which then turned the oxides into metals for use in the various applications. The mine did have a small percentage of neodymium, but it was tossed aside: there was no use for it until the late 1980s, when Magnequench came looking.

While Molycorp slogged along, in the late 1980s, Chinese geologists realized they had their own rare earths—an enormous deposit of monazite and bastnaesite intertwined with a monstrous iron ore deposit in Inner Mongolia. The Bayan Obo deposit dwarfed Mountain Pass a hundredfold, and China, at first unsure of how to process the complex minerals into metals, began a massive effort to turn their good fortune into pay dirt with the backing and momentum of Program 863.[16] By the 1990s China had flooded the markets, and prices for rare earths crashed. Small-time Chinese miners also discovered more than a hundred deposits of heavy rare earths in a type of highly weathered rock called laterite in southern China.[17]

Molycorp limped along for another decade against Chinese competition and, after a toxic spill, environmental regulators shut it down. For China's leaders, dominating the rare earth industry was a very important tactical maneuver. "Improve the development and application of rare earths," President Jiang Zemin wrote, "and change the resource advantage into economic superiority."[18] By the late 1990s, China had not only swallowed up Magnequench but, in less than a decade, captured the entire rare earth value chain, from mines to magnets.

19

China Inc.

If lawmakers had paid attention to Christopher Cox's alarming 1999 congressional report for the Select Committee on National Security and Military/Commercial Concerns with the People's Republic of China, they wouldn't have been surprised by China's rare earth monopoly at all. Capturing the rare earth value chain was just part of China's far-reaching strategy. The ultimate plan was to get a tight grip on the beating heart of American enterprise.

It worked like a dream.

Complaining of being overtaxed and overcontrolled back home, beginning in the early 1990s, one multinational corporation after another—not just from the United States but from all over the world—was quietly seduced to the Chinese mainland. The companies were lavished with government contracts and backstage access to the world's fastest-growing consumer market. Willing Casanovas like GE, Siemens, Kawasaki, and hundreds of others decided that for never-ending money in the land of plenty, they would do anything—including selling their company's intellectual soul.

The liaison was great for a while. China bought products from American, European, and Japanese companies made in China. Destitute workers came from all over the country to colossal industrial centers in the South. More than 170 Fortune 500 companies set up shop in the city of Guangzho, just a hundred miles up the Zhujiang River from Hong Kong.[1]

The tsunami of science and tech companies rolling into China was not missed by U.S. State Department analysts in Beijing. In 2005 the U.S. Embassy sent a report to Congress titled *U.S. and China Science and Technology Cooperation*:

> China has many alternative means of acquiring military and industrial technologies that are of far greater significance than the S&T Agreement [an agreement for cooperative science exchange signed in 1979 by U.S. president Jimmy Carter and Chinese president Deng Xiaoping]. Among these is a network of Chinese agreements for military, commercial, industrial and S&T cooperation with virtually every advanced industrial nation. Many of the latter are eager to sell, license, share or otherwise transfer their technologies to China. The lure of China's potentially gigantic consumer market also attracts massive private foreign investment from around the globe. Many U.S. firms are among those seeking to establish footholds in the Chinese market.

The State Department cable fell on deaf ears in Washington, D.C., where a congress full of professional politicians, lawyers, and businessmen preached Milton Friedman's free-market ideology and continued paving the way for one company after another to leave the United States and begin manufacturing high-tech products in China.

It was a dreamy coupling. American and foreign companies got richer. Chinese companies got smarter. But as in any relationship, over time things began to change. By 2006, Deng Xiaoping's brilliant idea to invite foreigners to make and sell their wares in China in exchange for company secrets began to feel stale and so last-century. So China made the strategic decision to put Chinese companies first. Why should foreigners get all the choice government contracts? Why should Western companies suck up the massive share of consumer wealth, which was growing at an unprecedented rate?

One of the highest-profile beneficiaries of the new China Inc. model was the son of President Jiang Zemin. Jiang Mianheng had returned to Shanghai from the United States in 1992 with his doctorate in electrical engineering and a stint at Hewlett-Packard. It was a grand time to be back in China. Like Deng Xiaoping's daughters and their husbands, who'd taken over Magnequench, Jiang Mianheng was part of a power clique of major Party leaders' children turned high-profile executives. Common Chinese people and foreign media often referred to the elite progeny, often derogatorily, as princelings and princesses. Young and highly educated, they moved easily on a global stage.

Not only was Jiang Mianheng an excellent researcher, with a serious curriculum vitae and a passion for science, but he knew very well that there was immense profitability in technology. Apparently against his father's wishes, he wasted no time jumping into the business world. He joined the Shanghai Institute of Metallurgy, where he spun off a for-profit technology company. Two years later, he started an investment company in Shanghai to take scientific discoveries into the commercial realm. He was then appointed to the influential post of vice president of the prestigious Chinese Academy of Sciences. Jiang's dream, he told friends, was to develop a self-sustainable model for scientific research in China, where basic science was funded by the end products it created. He called for national research programs in every field: alternative energy, coal liquefaction, electric cars, mobile phone networks, particle accelerators, spaceships, and lunar satellites.

By the end of the 1990s Jiang had recognized that China had an immediate need to be self-sufficient in all technologies, especially the most important component, the silicon chip or semiconductor. American chipmakers Intel and IBM were the only game globally, so Jiang and Wilson Wong, son of a Taiwanese business tycoon, started a business called Grace Semiconductor Manufacturing Corporation (GSMC).[2] The company initially had trouble with quality and fabrication. They needed permits to import sensitive equipment from the United States to etch the ultra-thin silicon wafers. So in 2002, to ramp up the political clout of their company in the United States, Jiang and Wong found themselves an American princeling of their own to serve on Grace's corporate board. With zero tech know-how and a history of running investment scams,[3] Neil Bush—son of former president George H. W. Bush and brother of then U.S. president George W. Bush—was nominated to the board to "provide GSMC from time to time with business strategies and policies; latest information and trends of the related industry, and other expertise advices." According to his contract, which was made public in Neil Bush's messy 2002 divorce, he was offered $400,000 a year over five years.[4]

Having already made business connections at IBM in California, Jiang was soon courted by foreign investors who saw him as the country's most valuable business partner. When he launched China Netcom in 1999, Rupert Murdoch and Goldman Sachs came in with $325 million.[5] The start-up was drowning until Jiang's father broke up China's biggest telecom company and gave one-third of the market to his son's company. Jiang met with GE's Jack Welch, Sumner Redstone of Viacom, and Yahoo founder Jerry Yang, who told the *Wall Street Journal* that Jiang was "very open to working with Western companies."[6]

There were cries of nepotism. Some called Jiang a princeling, reaping the benefits of his political connections. But there was no denying how well suited he was to the role of business mogul with an eye on China's future. Jiang's telecom empire exploded into the twenty-first century, taking advantage of China's self-serving new economic strategy. At the same time, foreign industries felt the fallout of the new China Inc. model almost immediately. Many felt blindsided as their profits slipped. By 2010, 31 percent of American companies doing business in China had complained to the American Chamber of Commerce in Beijing that they were being squeezed out of lucrative government contracts.[7]

A disgruntled China devotee named Jeffrey Immelt was beside himself. After doing cartwheels to get a grip on the Chinese market, giving up all kinds of patented technology, from rail locomotives to antipollution equipment to avionics, the bombastic CEO of General Electric complained at a conference in Italy. "I really worry about China," he warned his Italian contemporaries. "I am not sure that in the end they want any of us to win, or any of us to be successful."[8] In 2010 GE did barely half of the $10 billion worth of business in China that Immelt had predicted. Perhaps Immelt was exaggerating GE's pain; the following year, after being appointed President Obama's "jobs czar" for America, he shut down the last remnants of GE's medical equipment division in Wisconsin and sent it to Beijing.

Other companies cried foul. After scoring a contract for high-speed trains, Germany's Siemens was pushed out of future deals as soon as China's new railway company had gobbled up the company's intellectual property. Same thing happened for a Dutch company making wind turbines. Microsoft's CEO complained that the Chinese government was pirating their software, and then winced when leaders in Beijing said they'd make good by giving Microsoft 10 percent of the actual software cost. FedEx, DHL, and UPS all got squeezed by China's new and improved China Post, which mirrored the foreign carriers' tech know-how.

After decades of telling the U.S. government to stay out of their business, the heads of nineteen petulant corporate lobbying groups went to Washington, demanding something be done about China. "The AmCham China [American Chamber of Commerce in the People's Republic of China] survey shows that U.S. companies believe they face product discrimination in state-owned enterprise purchases, as well as in government procurement," the business leaders wrote.[9] *China is not playing fair. China is helping Chinese companies at our expense. They took all our good intellectual property, and now they won't give us our fair share of their $85 billion in government contracts.* Commerce secretary Gary Locke rushed to the rescue: "We recognize that this issue is just one facet of a broader Chinese approach to industrial policy that is creating headaches for U.S. companies operating in and trying to export to China. Successfully dealing with this will require both tactical and strategic adjustments to our engagement with China, and I can assure you that the administration is diligently studying what steps might be necessary."[10]

China ignored the veiled threats and grievances. Perhaps Chinese leaders found it amusing to watch the world's biggest multinationals squirm after being so quick to barter off their secrets. Or perhaps they saw irony when U.S. leaders—who'd killed trade regulations that prevented corporations from offshoring all things American—were irked that China wanted to protect its own. Either way, China had little reason to worry. Their new economic strategy—based on what economists call mercantilism—worked swimmingly. In 2010, while most of the globe was still reeling from the worst global financial disaster ever, China's economy bounced back to double-digit growth. China became the world's Walmart, exporting everything everywhere at bargain basement prices. America was its biggest customer, and ended 2010 with a $273 billion trade deficit.

The enormous revenue stream from state-owned businesses and corporate taxes allowed for an unparalleled Chinese government spending spree. Vast funds went into research and development for technological innovations. Foreign companies, real estate, and natural resources were snapped up in bulk all over the globe. The People's Liberation Army was one of the biggest winners, scoring state-of-the-art nuclear ballistic and cruise missiles, cyberattack capabilities, naval ships, aircraft, and weaponry. A new Chinese-designed missile tested in 2007 took out a weather satellite at 537 miles. While China claimed that it was all in the interest of peace and national defense, the U.S. military brass was not buying it. Admiral Robert F. Willard, head of the U.S. Pacific Command, warned the House Armed Services Committee that "the PRC's stated goals of a defense-oriented military capability contributing to a 'peaceful and harmonious' Asia appear incompatible with the extent of sophisticated weaponry China produces today."

20

Smoke and Mirrors

At every point I have endeavored to collect reliable statistics of our mineral wealth. The numbers of those given might have been considerably increased, but I preferred to under rather than overrate the produce of our mines, knowing with how little discrimination estimates are often made, and fully conscious that nothing, at home and abroad, is so injurious to mining interests as false and exaggerated estimates. Their influence is to create doubt and skepticism; or, if perchance they should gain credence for a time, it is almost invariably to the great detriment of individuals, and ultimately, by reaction in public confidence, to the great injury of all mining interests.

—GEORGE CLINTON SWALLOW, MISSOURI STATE GEOLOGIST, 1854[1]

More than a decade after China captured the rare earth market, in March 2010, Mark Smith arrived in Washington, D.C., and swore he would save the day. The former Unocal Mining executive testified to a congressional subcommittee that his new company, Molycorp Inc., would disengage America from the dangerous Chinese monopoly.[2] To make it happen, he needed federally guaranteed loans, he said. When Pennsylvania congresswoman Kathleen Dahlkemper pressed Smith on his plan to ensure Molycorp wasn't the next Magnequench—something to be uprooted by China—Smith gave an impassioned and patriotic response. "The odds of another Magnequench happening as a result of investment in our project are practically zero," he testified. "That is because we, as a company, believe very strongly that America's national and economic security demand that the entire rare earth magnet supply chain be established and maintained here in America. This is our goal and our commitment."

Smith left Washington without government-secured loans to reopen Mountain Pass—an endeavor for which he needed nearly half a billion dollars. But with prices of rare earth metals exploding due to China's trade limits, Molycorp's ship came in just four months later. On July 29, Smith pulled off the second most successful initial public offering (IPO) of 2010, raising more than $393.8 million.[3] "We are the only rare earth oxide, or REO, producer in the Western Hemisphere and own one of the world's largest, most fully developed rare earth projects outside of China," the Securities and Exchange Commission filing bragged to investors.[4] "Global demand for rare earth elements, or REEs, is projected to steadily increase due to continuing growth in existing applications and increased innovation and development of new end uses."

Lawmakers in Washington breathed a collective sigh of relief. Mark Smith was their white knight, and Molycorp—with its promise of bountiful rare earths—was his shining armor. The rare earth problem would be solved. The Pentagon could decouple from China. Manufacturers still making things in the United States could access rare earth components—all with zero help from the U.S. government. The market would succeed as free-market capitalism should.

Not everyone was waving an American flag over Molycorp's meteoric debut. "Molycorp's entire business plan consisted of just three letters: I, P, and O," says Jim Kennedy. "The fictional story was front and center on their own SEC documents, but no one bothered to evaluate the mineral and metallurgical information. No one bothered with the due diligence beyond the financial projections." Jim knew that, technically speaking, Molycorp did have economic levels of neodymium, but they did not have any of the other critical heavy rare earths necessary to sustain the modern economy or the defense industry. With the deposit containing less than one hundredth of a percent europium—a light rare earth used in color monitors and lasers—the company proudly announced that europium would be one of its major products.[5] On top of it, the company claimed to have dysprosium (used in neodymium magnets to make them viable at even higher temperatures) and terbium (used in the production of

Terfenol-D, a part of the U.S. Navy's most advanced sonar systems), though these two elements had never been found in economic concentrations in the mine's fifty-year history.[6] In fact, Molycorp's SEC documents clearly demonstrated that the deposit did not contain economic levels for half the rare earths needed to eliminate U.S. dependence on China. Molycorp was a fairy tale, Jim believed, and the only ones who would see a happy ending were those cashing in on the $393.8 million IPO.

With Molycorp's success, rare earth elements became the flavor of the month. First dozens, then hundreds of rare earth start-ups went looking for investors. "Many of these projects were led by unscrupulous or ill-informed promoters imagining their project would be the next Molycorp IPO," says Jim. Some were actually excellent deposits of heavy rare earths, but as with Pea Ridge, they were plagued with thorium. Jim knew that none of the mines could compete with the massive amount of resources that China had on tap.

The larger issue making Molycorp a ridiculous concept was something that Jim learned during exhaustive research. It turned out that in reality, the United States and many Western nations were literally drowning in rare earth resources. What Pea Ridge had in rare earths two thousand feet underground was just a drop in the bucket of what was sitting in enormous piles of mining waste on the earth's surface in the United States and countries all over the globe. Monazite, xenotime, and apatite (three of the main rare earth ores) are "byproducts" of existing mining operations across the world. What Jim owned at Pea Ridge in his breccia pipes and tailings—in fact what any stand-alone rare earth mine project had in its reserves—was dwarfed by the leftovers from iron, phosphorus, and titanium and mineral sand mining.[7] "If the thorium problem went away, the entire world would be flush with bankable rare earth elements," says Jim. No one needed to go to the expense of starting up a mine for rare earths, and no investor should be bilked into doing so.

"The more I discovered how abundant this resource was, the more I realized that at most, my mine would just be a small contributor," Jim says. In truth, that really didn't matter to him. By that time, he "understood the geopolitical ramifications of the problem. I was fully engaged with the U.S. government trying to fix it. The solution was bigger than me and the

Pea Ridge mine." Jim thought of his rare earth cooperative idea as a Manhattan Project of materials science. He still wanted to be involved in something so incredibly important. He planned to be a sort of director of the co-op. He would help get rare earth producers involved. The producers would deliver ore that the cooperative would convert into oxides, then metals, then end products. More importantly, he wanted to bring in multinational corporations—the end users—as partners. Jim saw the cooperative as an integral part of the United States economy—even more important than Silicon Valley, he says, "because it's a physical product that's the basis for the future of our manufacturing economy. It's not just one more app."

Still, Jim's cooperative idea was a hard sell. In February 2011, many lawmakers were sure the United States was home free when Molycorp became "one of the fastest windfalls in private-equity history," according to the *Wall Street Journal*, "turning $200 million into a paper profit of about $2.3 billion in just 30 months, or roughly $2.6 million in profit each day." In less than twelve months, the share price reached $75—or nearly $7 billion in market capitalization. Molycorp had everyone's attention.

Molycorp's well-oiled publicity machine, led by Vice President of Corporate Communications Jim Sims, continued to tout the company's quest to singlehandedly untangle America from the Chinese monopoly. The tune they whistled over and over at conferences, at legislative hearings, and in the media reverberated across the marketplace. At the time, there was only one actual competitor to Molycorp outside China—an Australian company called Lynas with a mine in western Australia called Mount Weld. Lynas raised $450 million from JP Morgan in 2009 and another $250 million from the Japanese government, who were desperate for a stable source of rare earths.[8] Significant problems plagued the operation from the get-go. Because the ore contained thorium and uranium—and Australia has some of the world's most restrictive laws regarding radioactive materials—Lynas shipped it off to be processed in Kuantan, Malaysia. A huge backlash ensued when the Malaysians decided they wanted nothing to do with the toxic processing plant in their town.[9] Thousands came out, month after month. Some speculated that China was behind the backlash. But it was the free market that was quickly sinking Lynas and Japan's hope for a secure source of rare earths; against the downward-inching Chinese-set price, Lynas's product was only worth half of the cost to make it.

With Jim's life in turmoil back in St. Louis, he and John Kutsch went to Washington, D.C., again and again. They went to universities, think tanks, and corporations—anywhere they felt they could find people to help them with their plan.

"Jim was in a really tough way when we were doing a lot of lobbying," says John. Since Jim had no income at all, his assets were frozen, and he was millions of dollars in debt, John covered all of their travel expenses. "He was in a constant state of panic and fear. I really felt horrible for what he was going through." For Jim, his work away from home on Capitol Hill was a distraction as Pea Ridge slipped away from him. It helped him cope. In the offices of U.S. lawmakers or Pentagon officials, Jim's problems suddenly seemed small in contrast to what was happing to his country. "The entire manufacturing economy was pulled out from under the United States by forces of greed so anaesthetizing that almost no one even saw it coming," he says. "No one was paying attention."

From their previous forays into the national political scene, Jim and John knew a little about how Washington, D.C., works. They did not expect to get direct meetings with lawmakers, because in the halls of Congress, all business goes through legislative assistants, informally called staffers. Staffers deal with specific topics for their bosses, who can be senators, representatives, or committees. So Jim and John targeted staffers who worked on energy and defense concerns.

In wood-paneled offices on Capitol Hill, Jim excitedly encapsulated everything he'd learned in the past seven years into fifteen-minute briefings. He explained that no single company could break the Chinese monopoly. He hurriedly tried to unravel the scientific jargon of rare earths, costs and market manipulations, upstream and downstream values. He explained that rare earth oxides—what Molycorp and all the other mining projects were offering—were worthless to the defense industry and technology companies without the entire value stream in place. You can't take neodymium oxide and make a smart bomb. You need magnets, and those still come from China too, he would say. John chimed in to explain the thorium molten salt reactor, and how thorium from rare earth deposits could power a fleet of safe nuclear energy. Jim and John had a significant disadvantage in that they were not versed in Washington-speak. They didn't know how to condense their very complicated arguments, and sometimes Jim got crass and defensive when people didn't get his message—a serious D.C. no-no. Lots of times, they got blank looks and aides checking their clocks. "The staffers just gave up. They're like, 'Heavy? Light? What's the difference?'" Jim honestly couldn't understand why it was so hard to get his message through. "You never get past the real basics with most of these people in Washington," he says.

Occasionally, though, Jim and John would have some success with a staffer. On one of their many trips to D.C., John left in the morning to speak at a nuclear energy conference. With an hour to kill, Jim made a call to Senate majority leader Harry Reid's office and asked to speak with the energy staffer. Jim heard that Reid had a bill with Utah senator Orrin Hatch called the Thorium Energy Security Act of 2010, which called for new regulations that would allow thorium fuel in nuclear reactors. Amazingly, the staffer invited him over. Reid's staffer was gung-ho on Jim's idea, and sent him to Hatch's energy staffer next. "I couldn't believe it. Someone was very interested, and it was the *majority leader's* office. I frantically called John and said, 'Get your ass over here now!'"

In his glass cubicle equipped with wall-to-wall computer screens, Hatch's energy staffer, J. J. Brown, was cordial, bright, and knew a lot of key people in the nuclear business. He explained how he came to work on the new thorium bill: A small energy company came to Senator Hatch, looking to add thorium to traditional nuclear fuel. The company was testing in Russia, and the owners wanted to bring the technology to the United States, but the regulatory and testing process was slow and expensive, and the company met resistance from the uranium mining and fuel industry. Brown did some investigating and found that the National Nuclear Security Administration (NNSA—a subagency of the Department of Energy) was hiding a report by Westinghouse by making it "classified." Senator Hatch forced the report out, and Brown looked it over. To Brown, it wasn't classified information at all. "The report said that Westinghouse had looked hard at thorium, and the conclusion was that thorium was something that ought to be pursued," says Brown. "I think NNSA sat on [the report] because anything new is sometimes a threat to the established guys. They don't want something a thousand times better coming along and destroying all their investments."

Brown immediately liked John and Jim and their Midwestern approach. "They came across as not very slick in terms of not speaking Washington lingo," he says. But unlike a lot of staffers, Brown enjoyed taking meetings with folks who didn't come accompanied by professional lobbyists. It was harder to dissect their arguments, but he enjoyed the challenge. Brown instantly picked up on Jim's and John's passion and found them highly intelligent. John captured Brown's imagination with the story of the molten salt reactor. It was an intriguing energy concept. "The simplicity of it, the safety of it, and the fact that it had been tested at Oak Ridge for so long made me think my bill also needed to include the molten salt reactor."

During the meeting, Jim brought up the subject of rare earth elements and the Chinese monopoly, but Brown's brain was spinning with thorium nuclear energy. Rare earths were not on his radar at the time. For Jim and John, thorium and rare earths are intrinsically related, but selling that point was proving difficult in Washington, where neither thorium nor rare earths had grabbed lawmakers' attention.

Sadly—especially for John—the thorium bill fell by the wayside. But by 2011, after the fishing boat incident, lawmakers were suddenly wide awake to the topic of rare earth elements. Multinational corporations, desperate to get the Chinese quotas lifted, circled Capitol Hill. China was part of the World Trade Organization, having joined in 2001. Since the WTO bars export quotas in most cases, defense contractors, medical manufacturers, and high-tech equipment companies went to Washington, complaining about China's noncompliance. Most corporations that needed rare earths would not publically denounce China, lest they be subject to some sort of retaliation. But on September 21, 2011, John Galyen, president of Danfoss North America, an electronics manufacturing company in Baltimore, testified before the House Subcommittee on Asia and the Pacific: "For Danfoss, two elements—Dysprosium and Neodymium—are critical to production of high-efficiency variable speed compressors and the oil-free centrifugal compressors produced by Danfoss Turboco. As a result of the reduced export quota, manufacturers like Danfoss are experiencing an 800 percent increase in the price of magnets used in our compressors and protracted lead times, going from eight weeks to six months."[10] In his conclusions, Galyen asked that the United States file a claim in the WTO to pressure China to increase quotas.

To Jim, a WTO case forcing China to drop export quotas was the worst possible scenario. It would hand China a victory regardless of which way it went. Jim pulled every string he could with Missouri senator Claire McCaskill's office and got himself a meeting with the White House Office of Science and Technology Policy. Jim and John arrived at the office of Assistant Secretary Cyrus Wadia, a chemical engineer with a Ph.D. in energy and resources. "We told him that the WTO is a trap. China anticipated the move, and it will be disastrous for the United States," Jim says. "We tried to explain that if the WTO forced China's hand, China would release a flood of low-value rare earths, driving prices lower and bankrupting Molycorp and others." Wadia's office was in the Eisenhower Executive Office Building, the same place where Alvin Weinberg, father of the thorium molten salt reactor, spent a miserable year after being fired from Oak Ridge in 1972. When they told Wadia about Weinberg's TMSR, he seemed intrigued. "We asked him if he could do something. Then he summed up the U.S. energy policy with two words," says Jim, "like a Facebook relationship: 'It's complicated.'"

Across town in the Russell Senate Office Building, Jim had more luck. Roy Blunt, the junior senator from Missouri, took a serious interest in Jim's rare earth cooperative idea. Jim's excitement grew when Blunt's senior energy staffer said they would find a Democratic cosponsor and introduce the National Rare Earth Cooperative Act in the 113th Congress. Jim and John were euphoric. "We were going to change the world for the better," says John. "We were going to make the USA strong again. We had done the impossible, and done it without breaking a sweat."

What happened next was a blow that they were completely unprepared for. As Jim, John, and Blunt's staffer worked on getting support for the bill, the very agency that the bill was designed to help made it clear they didn't need any help at all. The Pentagon's long-awaited report to Congress, "Rare Earth Materials in Defense Applications," went public in March 2012.[11]

The five-page analysis, which had taken two years to complete, blindsided Jim. "Rare earth materials are widely used within the defense industrial base; however, such end uses represent a small fraction of U.S. consumption. As a result, when looked at in isolation, the growing U.S. supply of these materials is increasingly capable of meeting the consumption of the defense industrial base." The report, signed off on by Undersecretary of Defense for Acquisitions Frank Kendall, claimed that by 2013 the problem of the Chinese monopoly would be mostly solved by "free-market" domestic production—with a little help from the government to find "substitutes" for rare earth elements. By 2015 there would be nothing to worry about at all, the report said. The media excitedly picked up on the end of the crisis. "Rare Earth Supplies in U.S. to Meet Defense Needs, Pentagon Says," Bloomberg reported.[12] "Pentagon Plays Down China's Rare-Earths Controls," declared the *Wall Street Journal*.[13]

Jim's blood pressure soared as Washington think tanks and policy shops chimed in and wrote papers lauding the free-market solution and Molycorp—the company that had brought rare earths back to America. Congressional Research Service analyst Valerie Grasso in particular gained Jim's ire for her 2012 report to Congress arguing that Molycorp was poised to help solve the rare earth crisis. "It sounded like a press release for Molycorp," he says. That wasn't surprising, considering that one of Grasso's primary sources of information was Molycorp's assistant public affairs officer.[14] "Molycorp was capitalizing on the success of its spin machine, making glaringly false statements about its capabilities and resources, including magical powers of alchemy allowing it to produce heavy rare earth elements from nothing," says Jim. Paradoxically, in the same report, Grasso listed a U.S. Geological Survey study showing zero heavy rare earths at Mountain Pass.[15]

Jim fired off a long and testy e-mail to Grasso and then to every senator and Congress member on the Energy and Defense Committees. "The conclusions of [the DoD] report cannot be supported by Molycorp's own documents. This is further confirmed by extensive USGS [U.S. Geological Survey] documentation and the 50-year operating history of the mine . . . If you understand the rare earth market in any detail you will see that China will open the flood gates on La (Lanthanum) and Ce (Cerium) once the 'West' starts up production . . . This will crash all of these companies, making China even more powerful than before."

The other DoD claim that made Jim furious was the idea of "substitutes" for rare earth elements. Even though Molycorp had a bulk of lawmakers convinced they had everything under control, a still-concerned Congress stepped off the free-market train to the tune of $120 million over five years for a new Critical Minerals Institute.[16] Led by the Ames Laboratory, government, university, and industry partners were charged to find substitutes for rare earth elements in high-tech applications. One of the corporate partners was Molycorp. To Jim, it was just another stupid smokescreen. "What were they looking for," Jim asks, "some yet undiscovered elements lurking in the shadows of the periodic table? Nothing known has the same magnetic properties as rare earths." Any new technologies could take decades to develop. Then military systems would have to be redesigned and retrofitted to work with new parts. The scientists at the national labs knew this, Jim was sure, "but desperate for government funding and a desire to remain relevant, they took the money and ran."

For Jim, it all felt like the plot of a seedy spy novel—*A powerful government falls victim to deeply flawed intelligence that could easily be vetted and dismissed as preposterous.* "The Pentagon and Congress were playing hapless dupes in this drama," says Jim. Even if Molycorp was able to produce some dysprosium oxide through an act of alchemy, then what? There was nowhere to process it into a metal outside China; no way to turn it into an alloy or a magnet.

There was another person in Washington also not buying the DoD's rosy take on rare earths. Twenty years had passed since Peter Leitner tried to stop Magnequench from leaving the United States for China. Toward the end of his career as the senior strategic affairs adviser for the Pentagon, he saw the rare earth crisis unfold. To Leitner, it was an epic fail by his government and his former employer. Like Jim, he believed that it was his duty to do something to stop China from gaining a 100 percent monopoly. In 2005 the Chinese National Oil Company made a bid for Unocal, the parent company of Molycorp. "I was convinced if they got hold of Molycorp, they would plug up the mine and donate it to California as a nature reserve. I was trying to actively stop it." Leitner generated enough discontent on Capitol Hill that "several congressmen issued a letter saying that if CNOC attempts to take over Unocal, they will block it." The Chinese backed off. Leitner continued to call out the dangerous and broken system he saw in the Pentagon. In 2008, a retired Leitner told *Forward Online* that the REE situation is "a form of economic warfare, and no one is really paying a whole lot of attention to what the Chinese are doing. [The Pentagon leaders] are embarrassed by the fact that they don't understand their own acquisition process. We are totally at the mercy of contractors, subcontractors and sub-sub-subcontractors . . . When it comes down to actually understanding where our dependencies are and whether we can confront them, those are issues that nobody at the Pentagon really wants to face."

In 2008, high-ranking Pentagon officials testified to Congress that they weren't buying anything made in China at all—even as nearly every one of their systems contained Chinese-sourced parts. "What more can an individual do?" Leitner asks. "I did more than anybody else in the government as far as I'm concerned—including jeopardizing my career." As vocal as he had been, Leitner was never forced out of the Pentagon. "I left because the mission was abandoned by those in charge," he says.

A lethal blow to those still trying to combat the Chinese monopoly came on March 16, 2012, delivered by the president of the United States. From the White House Rose Garden Barack Obama, backed by Japan, the European Union, and a multitude of business interests, was about to force China to the punch. "This morning, we're taking an additional step forward. We're bringing a new trade case against China, and we're being joined by Japan and some of our European allies," Obama said to reporters at a gathered press conference.[17] "This case involves something called rare earth materials, which are used by American manufacturers to make high-tech products like advanced batteries that power everything from hybrid cars to cell phones. We want our companies building those products right here in America. But to do that, American manufacturers need to have access to rare earth materials—which China supplies. Now, if China would simply let the market work on its own, we'd have no objections. But their policies currently are preventing that from happening. And they go against the very rules that China agreed to follow."

China would feign resistance to the WTO case, but it didn't matter. They dropped the quotas. The market flooded. Prices tanked. And like a row of dominoes, rare earth projects fell one by one—every single one before a shovel even broke ground. Only Molycorp was left standing, albeit shakily and with its stock prices tumbling. Still, Mark Smith continued to tout his company's exclusive status as a homegrown supplier of rare earth oxides, metals, and magnets. Meanwhile, defense contractors continued about their business of ordering Chinese components for American defense systems. There were laws preventing them from doing so without a complicated waiver process, but it was rarely used, and the Pentagon turned a blind eye.

With Molycorp hemorrhaging cash, in March 2012 CEO Mark Smith made a curiously bold move. He borrowed $1.3 billion in capital and bought a rare earth processing and magnet company called Neo Material Technologies that was registered in Canada.[18] Smith had promised "mines to magnets"; now he had all the components to deliver. Still, Molycorp stock was in the toilet, and the company was not yet in production after nearly two years. To Jim, something smelled foul about the Neo Material deal. But Smith was like a free-market god in the world of Washington, D.C., and Jim—who kept yelling that it was all a big fraud—was Chicken Little. Exasperated staffers took a deep breath when they saw him coming. *Go away, Jim. The sky is no longer falling. Molycorp has got this.*

At the beginning of the 113th Congress in 2013, U.S. senator Roy Blunt was about to seek out a Democratic cosponsor and offer up the National Rare Earth Cooperative Act for a vote. Then Blunt got wind of a lawsuit involving Jim Kennedy and a powerful corporate constituent in his home state, Alberici Constructors. Blunt dropped the bill—a normal and sensible move for a lawmaker, according to J. J. Brown, but a soul-crushing blow to Jim Kennedy and John Kutsch.

Back in St. Louis and at the end of his rope, Jim went to see his old friend Rich Todaro. Jim often went to Rich to vent. Rich was solid, calm, and had none of Jim's hot-headedness. Jim ranted and rambled in techno talk and policy jargon punctuated by acronyms and expletives. Rich barely followed. "I don't know all the details about why he didn't like Molycorp," he says. "But in general, I would tell him, 'Jim, if you would work with everyone in the industry, you could get everyone working toward the same thing.' My counsel was, 'Don't make enemies.'" Jim would never take Todaro's advice. "This was an ongoing discussion that began well before Molycorp IPOed," Jim says. "Rich would tell me, 'Align yourself with Molycorp.' In short, he was telling me to do what everyone else was doing: Screw the USA. Play along, and make some money." Todaro wasn't surprised that Jim didn't want his advice. It was always the same. "In Jim's mind, he believed Molycorp was doing something wrong," Todaro remembers. "They were telling someone the wrong story. So even if it's to his detriment, he says, 'I don't care. I'm going to do what's right.'"

21

The First to Eat a Crab

As the first decade of the twenty-first century came to an end, Jiang Mianheng—now in his mid-fifties—was one of China's most successful businessmen. He was known as the Telecom King, a titan in a new global economy with China front and center. His interests were varied, and his portfolio boasted a wide variety of endeavors. With the fund he ran for the Shanghai government, Jiang bought a stake in Oriental DreamWorks, the Bank of Shanghai, and China Eastern Airlines. He joined ventures with Microsoft, Nokia, and MetLife China.[1] He was decidedly unflashy, and he refused interviews—common practice for high-level Chinese. Some said that his father, former president Jiang Zemin, who left office in 2003, wished his son would stay in academia, pursuing scientific endeavors, and out of the business sphere. But Jiang seemed completely confident living in both worlds, and using each to benefit the other. Even with access to the very highest echelons of power and the funds to partner in any industry he chose, though, there was one very important thing he wanted—something that he couldn't yet get his hands on.

China's exploding energy needs over the last few decades were met mainly by burning dirty coal in antiquated power plants. The pollution in China's big cities—especially Beijing—was a global disgrace and caused a health epidemic in the population. The obvious option to wean China off coal was nuclear. China had planned to develop "indigenous" reactors at the turn of the century, but in 2006 they ditched the homespun idea and decided instead to engage foreign suppliers. Westinghouse—at the time a British-owned American subsidiary—won a $5 billion contract to build four Gen 3 (third-generation) reactors—the most advanced model of light-water reactors, though the advances were in safety features, not core design changes.[2] To seal the deal, Westinghouse agreed to hand over all the engineering plans for the new AP1000. About this part of the contract, the Chinese government was "very demanding," according to the U.S. secretary of energy. Westinghouse hadn't built a new reactor since 1996, and some said the company gave up all of its intellectual property, expecting it would get future consulting gigs. The company also hoped to show U.S. customers that it had a new and viable reactor design. Unfortunately for Westinghouse, the AP1000 experienced many problems in its construction phase, and even before the first one was up and running, China's CPR14000—essentially a copycat of the Westinghouse model—was set to be the industry replacement in China.[3]

Jiang Mianheng was likely unimpressed by the CPR14000. The Gen 3 model still needed highly pressurized water to cool the core, so it required an enormous containment vessel and mammoth water supply. It still burned the same old solid fuel of enriched uranium—something of which China did not have plentiful reserves. In principle, the new Gen 3 reactor remained a Model T Ford refurbished with seat belts and airbags. China was moving at breakneck speed into the twenty-first century, and old-school nukes were not going to cut it anymore.

Fortunately for the Chinese government, they now had the means and the fortitude to embark on a nuclear energy revolution of their own. In 2000 they announced the creation of China Haiyan Nuclear Power City. With a ten-year budget of $28 billion, the massive industrial center would develop new nuclear industries, train and educate nuclear scientists and engineers, and work in applied sciences. It was just one of many well-funded nuclear initiatives.[4]

Jiang Mianheng, as the director of the Shanghai branch of the Chinese Academy of Sciences, would be in charge of two reactor projects. For his part, Jiang wanted something different from a souped-up old-school reactor. Jiang took the growing energy needs of his country very seriously, calling China's lack of indigenous energy a "security issue." To fix the problem, he said—speaking English to a conference audience—"We can rely on outside China, or we can develop ourselves." And he was about to do just that. Jiang was the newly named beneficiary of start-up funds from the People's Republic of China to the tune of $350 million.[5] Now he had the means; he just wasn't sure of the way.

That changed in the summer of 2010, when Jiang read an article by thorium advocate Robert Hargraves and physicist Ralph Moir in *American Scientist*.[6] The story was a history lesson as much as an article about nuclear science. America made a discovery in the 1960s in the field of nuclear energy—an idea that was radical and new. In a lot of ways, the design for the incipient reactor at Oak Ridge, built with public funds through the Atomic Energy Commission for the U.S. Air Force, looked far better than what took over the commercial nuclear industry. The design implemented thorium as a fuel, a material four times more abundant in the earth's crust and much easier to get than uranium. But the thorium molten salt reactor never emerged from the hills of Tennessee. Instead, because of a lethal mix of politics and economics, it was mothballed and forgotten. The operational logs, designs, blueprints, calculations, and notes, handwritten and typed, were in deep storage at Oak Ridge until 2008, when engineer Kirk Sorensen learned of the technology. With funding from NASA (the space agency was looking for new fuel sources for deep-space missions), Sorensen digitized everything he could find on the TMSR, and made it all public on an online forum at energyfromthorium.com. Soon a small group of engineers and physicists, including John Kutsch, was galvanized by the idea of bringing the project back to life.

The thorium molten salt reactor seemed like an excellent idea to Jiang as well. It didn't matter that it had never been done commercially. It didn't matter that the technology was old and many wrinkles still needed to be ironed out. It didn't matter that the United States had abandoned the technology in the early 1970s. The physics and the chemistry were solid. This he knew: China would be the first to have a molten salt reactor that would burn thorium, a fuel that China would not run out of for 20,000 years.[7] It could be placed everywhere in the country. It could be far from rivers or coastlines because it would not need water for coolant. It would take China into the future—ahead of the West with its century-old addiction to fossil fuels. "We have a dream," Jiang said at a 2012 conference.[8] "If we can produce clean electricity, we can drive our electric cars. We need something revolutionary to happen. After all, 'China is the first one to eat a crab,'" he said, referring to an ancient Chinese proverb that basically means "China will be the first to do what everyone else is afraid to."

A few months later, at the end of December, Jiang and a few of his colleagues from the Chinese National Academy of Science were on a plane to Tennessee. At Oak Ridge National Laboratory, Cecil Parks and Jess Gehin, two directors of the Nuclear Systems Division, warmly welcomed the visitors. The Americans were immediately impressed with Jiang's top-notch science background and deep knowledge of nuclear energy. The Oak Ridge bosses planned to take the Chinese team—which included a senior nuclear physicist named Xu Hongjie—around the campus to tour some of the flashier elements of Oak Ridge, like the gigantic $1.4 billion Spallation Neutron Source. But Jiang made it clear he was only interested in one thing. He told the Americans that he'd learned about thorium molten salt reactors. He'd heard that for molten salt research, Oak Ridge was the number-one place in the world.

So the group made a beeline to see what they could of the decommissioned Molten Salt Reactor Experiment. There was little left to see; the reactor was underground, and there was no way to get down into it. Then they went to see what amounted to the only contemporary molten salt research project at Oak Ridge. Against one wall of a giant laboratory was a contraption the size of two elevators side by side. One side was about ten feet tall, the other about twice that. David Holcomb, a blond, bearded nuclear engineer, explained how the steel pipes and pump were designed to test how molten salt worked in a system at different temperatures and pressures and with different chemistries. This "salt loop" was meant to be the first step to building a modern nuclear reactor cooled with molten salt rather than water. Unfortunately, the salt loop wasn't running. It was incomplete for lack of funding, and Holcomb couldn't demonstrate it for the visitors.

Jiang and his colleagues listened to Holcomb's presentation and looked over the salt loop. According to Chen Kun, one of the younger researchers on Jiang's team, the Chinese scientists had not been aware of how little research had been accomplished due to a lack of funds. But now it was blatantly apparent where the Americans stood financially. "They built this loop," says Chen, "but they didn't get money to run the loop to make experiments." The field trip may have been eye-opening to Jiang, whose budget for one reactor project was about five times the entire annual budget for advanced nuclear research for the U.S. Department of Energy's Office of Nuclear Energy. Jiang was, however, impressed by the deep intelligence of the senior Oak Ridge scientists, and was very interested in knowing more of what they knew. According to Holcomb, Jiang was also very up-front about what he wanted to do with their knowledge. "He plans to form a joint stock company," says Holcomb, "which basically means he wants to create a company and to build [thorium molten salt reactors], and list the company on the U.S. stock exchange because we have a better-regulated market."

As a general rule, most American scientists rarely spoke of the future commercial applications of their work, tending to put science in a higher realm than economics. This was not true for Chinese scientists. Since Deng Xiaoping's capitalist reforms, science and economics had been intimately intertwined in the People's Republic of China. "And the reason I don't think [Jiang] was shy about this is that they have 700 trained engineers. They have a huge head start on everyone else, and money to put into it," says Holcomb, who was a one-man show with a budget of zero in 2014. America, as a whole, was falling woefully behind in the race for new nuclear energy. In June 2014 the U.S. Government Accountability Office reported that even though the primary mission of the Office of Nuclear Energy is to "advance nuclear power through research, development and demonstration," an advanced prototype reactor was unlikely any time in the foreseeable future.[9]

China must have seemed like a research Shangri-La to the American scientists and engineers, who were mostly in their forties and fifties. For their entire careers, nuclear scientists, not just at Oak Ridge but in all of the national labs and many universities, struggled to remain relevant in a nuclear-adverse country. The purse strings for funding were held tight by the Department of Energy, and regulations were written only for the energy systems already in place. Even then, regulatory requirements by a DoE subagency, the Nuclear Regulatory Commission, were so intensive that new light-water reactors would take a decade or more to clear and cost more than $100 million in application and design fees. (For the

Westinghouse AP1000—a light-water reactor—the NRC process took twenty-five years.)

True innovation was impossible, because for something outside the status quo, there were no written regulations at all. So once something new was developed, there was no pathway for it to get out of a laboratory into commercial production. China faced none of these constraints. It was all about getting the newest and most creative projects to fruition, and fast. To Holcomb, the difference between China and the United States was staggering. "They're putting in production a hundred gigawatts of power every year for the next thirty years. It's absolutely an enormous amount. The entire U.S. nuclear fleet is the amount they're installing every year. The world has never seen this level of expansion."

Jiang asked if the Oak Ridge team would like to come on board with him. Perhaps they could work together? It was not an unprecedented ask. A science-sharing agreement had been in place since 1979 between the Department of Energy and the People's Republic of China, and there was collaboration on several high-energy physics projects.[10] Still, working on a nuclear reactor project was very different. "We said, we can't work with the Chinese government without the U.S. government permission," Holcomb says. Lab director Cecil Parks told Jiang that it wasn't up to Oak Ridge; any collaboration would have to be cleared at the highest levels of the Department of Energy.

A month later, Jim Kennedy and John Kutsch landed in Tennessee, fired up and sure they could get Oak Ridge excited about restarting the thorium molten salt reactor mothballed so long ago. After Jim's talk about his "thorium problem" and their hope-dashing meeting with Jess Gehin, Jim and John also toured David Holcomb's lab. They were less than impressed. The salt loop had cost Oak Ridge $4 million to build. John felt like he could have fabricated one himself for a tenth of the cost. He knew that Holcomb's hands were tied, and actually felt like Holcomb was a captive of the extreme bureaucracy of the national lab. "Dave and his staff are very good," John says. "Imagine if they had some support."

Jim Kennedy left Oak Ridge for St. Louis with a hole in his gut. Although his journey started with a mysterious file from Pea Ridge, he now firmly believed that rare earths were just a small part of a much bigger problem. As he dug deeper into the topic of America's energy dominance around the world, he came to believe that reviving thorium nuclear energy was much bigger and more important than the Chinese rare earth monopoly. Not only could the TMSR power America's electricity needs, Jim believed it could guarantee the United States 100 percent energy independence.

Studying military and economic history, Jim learned of a process to turn coal into liquid fuel, patented by German chemist Friedrich Bergius. The technique of "coal liquefaction" powered the German army through World War I and World War II. South Africa's leading source of liquid fuel has come from coal since the 1950s. China, with massive coal reserves and a history of using the liquefaction process with energy from coal power plants, is on record as researching ways to use nuclear energy to turn coal to liquid fuel. North American coal reserves contain four times the available energy in all known oil reserves in the world. The problem is that it takes three times the energy from coal to produce one liquid fuel energy equivalent. But using nuclear energy, Jim says, "Suddenly—you can make not just the United States, but any country in the world, independent from petroleum forever. That's the end of wars over oil." Jim felt like grabbing the United States by its shoulders to shake it awake. "If China realizes that new, clean energy sources are a matter of national security, and the Chinese government knows that controlling rare earths is paramount to controlling commercial and defense manufacturing, why doesn't anyone in America?"

Still, Jim wasn't about to quit trying to convince everyone he could that it all really did matter. Endless hours of e-mails, phone calls, and face-to-face meetings with congressional staffers in Washington paid off. In 2011 Jim's rare earth cooperative idea found the support of Missouri senator Roy Blunt, who planned on introducing the National Rare Earth Cooperative Act in the 113th Congress. The bill allowed for rare earth resources to become part of a multinational value chain based in the United States. It also included the plan that Jim Kennedy and John Kutsch had developed three years earlier. If the bill made it into law, it would not only solve the thorium problem; for Western and Japanese rare earth producers, the plan called for the development of thorium nuclear energy in America. Thorium from rare earth mining operations would be secured in a "thorium bank" while a newly chartered U.S. corporation developed a thorium commercial reactor. The language called for a governing board that included Oak Ridge National Lab, the Defense Advanced Research Projects Agency (DARPA), the Department of Energy, a member of the Thorium Energy Alliance, and a representative from big business.

Jim's jubilation about the pending bill was short-lived. He soon found that getting allies would be harder than he originally thought. When he called Jess Gehin, the director of nuclear research at Oak Ridge, to give him the good news, Gehin told him that Oak Ridge wanted nothing to do with the legislation. "He told me to remove ORNL from the bill." Jim refused. He was livid. "Oak Ridge originally developed, operated, and proved the technology, and I knew that this lent credibility to the legislation," he says. Afterward, Gehin called Senator Blunt's office and asked them to remove ORNL from the bill. Gehin says that he believed that it should be the DoE choosing the board, rather than Oak Ridge being specifically named in the legislation.

Later, Jim learned something he believed to be the reason for Gehin's reaction. Oak Ridge's dance card was already full. There was no room for a couple of Midwestern scallywags to influence the molten salt nuclear reactor business. After his visit to Oak Ridge in December, Jiang Mianheng had gone to the U.S. Department of Energy's Office of Nuclear Energy and asked if the Chinese Academy of Sciences could collaborate with Oak Ridge on molten salt reactors. The DoE accepted his invitation.

The deal between China, the Department of Energy, and Oak Ridge National Laboratory was not officially a secret, but there was no public release announcing the upcoming collaboration. However, one of the university researchers involved was not happy about it, and word got out. Jim was furious. He fired off a testy e-mail to his Department of Defense contact, Rick Lowden. Lowden, who was still employed by Oak Ridge, had been an ally of Jim's since their initial meeting in 2008. He'd even worked to get Jim's rare earth cooperative idea in front of those who could help make it happen. Now Jim wanted Lowden's help keeping China from getting the molten salt technology up and running before the United States did. "The U.S., the original developer of this technology, is conspicuously absent from the race," he wrote. "No surprise considering that ORNL and DoE have done everything possible to kill all U.S. efforts to develop TMSR while spoon-feeding everything they can to China. I sure would like to see the secret documents between China and ORNL (we know they exist). But then again, why rock the boat if the boat is already headed to China. History will not be kind to any of us."

Lowden was done with Jim Kennedy. He wrote back, "Mr. Kennedy; Please remove me from your distribution list. I am fairly tolerant, however, I have little respect for those who openly criticize others especially when one of the entities they chose to berate is where I work—ORNL." Jim could add Lowden to the long list of people he'd pissed off in Washington.

John and Jim couldn't get Oak Ridge or the Department of Energy to release any documents concerning the deal with China, so John convinced a lawyer named Robert Orr—a nuclear lover and member of John's Thorium Energy Alliance—to make a Freedom of Information Act request. Nine months later, on July 23, 2014, Orr received the documents. Peter Lyons, the DoE's assistant secretary for nuclear energy, and Jiang Mianheng, of the Chinese Academy of Science, signed the agreement between Oak Ridge and Jiang in October 2012.[11] The Chinese team wanted to work on thorium nuclear projects, but according to senior researcher Xu Hongjie, the DoE said they didn't have any programs working with thorium. So the two countries would work together on a molten salt cooling system for China's new Gen 4 reactor designs.

What Jim would learn next wasn't outlined in the memo. It was one of the hardest things to swallow in his fight to save rare earths and revive molten salt technology. Not only would China be collaborating with Oak Ridge, and—Jim believed—capturing all of the research done by Weinberg and his team and paid for by the U.S. government; the People's Republic of China would actually bankroll the U.S. Department of Energy to pay American scientists in American national laboratories to work for the Chinese.

"It's true," said Holcomb in a 2014 interview. "China will be paying my salary." In a completely unprecedented move, Jiang Mianheng would use several million of his $350 million budget to bankroll research at an American national laboratory and pay U.S. government researchers to work for China. "It's a trade," says Holcomb. "They're going to do a lot of testing that we simply don't have the money for, and they're going to get experienced people to help provide some direction." The Chinese, Holcomb said, have a lot of very young staff, "and we have senior people who can really help." Chen Kun (then a thirty-five-year-old Chinese nuclear physicist who acted as a liaison between the Chinese and Americans) says that the collaboration isn't exactly a partnership, because "the funding contributions are not equal." Why is the American government agreeing to take funding from the Chinese to develop something that American scientists created? "Assistant Secretary for Nuclear Energy Peter Lyons has the vision that molten salt has a future." Chen says. "But he's worried that since they can't support the project, it is going to be lost."[12] For China, helping to support Oak Ridge was easy, says Chen. "CAS [Chinese Academy of Science] funding shot up, and they have plenty of money from the central government." What do Oak Ridge and the U.S. Department of Energy get in return during the ten-year collaboration? Whatever the Americans find out, they will be allowed to publish in the open literature.

To David Holcomb, it didn't really seem like the best deal. "I think that the United States having its own program is really desirable. But that's not what the DoE has elected to do," he says. "The United States doesn't pay for research anymore in nuclear power." The total advanced reactor research budget in 2014 was $122 million, spread among several national labs and various projects.[13] Holcomb resolved himself. "China is going to do this anyway. We can either be part of this, and try to benefit from it, or not benefit from it," he says. "At the moment I do not see any signs that the U.S. government is going to elect to pursue large projects. But I don't know what will happen when some of these things start to become real in other countries. I'm certainly hopeful that we may have a Sputnik moment."

When it was revealed that Oak Ridge was working with China and giving the cold shoulder to Americans interested in thorium nuclear projects, Holcomb said the blogosphere of thorium advocates went ballistic. "You're selling out!" they said. Holcomb shrugs. "We don't have anything to sell, because everything we did was forty years ago, and we published it then."

John E. Kelly, deputy assistant secretary for nuclear reactor technologies at the Department of Energy, rejected the criticism. "Why are we interested in China today? We want to influence them. You can't influence by standing thousands of miles away and shaking your fists," Kelly says. "You have to get right up on them. If China doesn't do this, it will be incredibly harmful for the world. China has to get off carbon dioxide." Kelly also took exception to people being up in arms over the United States working with foreign scientists. "The thing is, our program was imported from overseas. Look at the facts—Einstein . . . Oppenheimer. It was an international program from the beginning."

Jim Kennedy, for whom patriotism runs through his veins like adrenaline, also pondered the question. Why would Holcomb, Gehin, and the Department of Energy work not only with China but *for* China? Where was *their* patriotism? Why didn't they fight to make a future of safe nuclear energy in America? "Energy is the most important thing of all," says Jim. "If we miss the boat on this, how can we possibly compete in the world economy? Are we just going to wait around and let China take something that we developed as a country fifty years ago, or are we going to find a way of challenging this?"

On one level, Jim believed that Oak Ridge scientists were collaborating with China as a matter of self-preservation in a world of shrinking budgets and possible pink slips. On a higher level, Jim realized that for some scientists, scientific research trumps nationality. "So when they realize the national labs have no funding, and that it will never happen here, and China's willing to do it . . . they may think, Maybe I'm screwing my country, but my allegiance is to science. They don't carry the guilt of the subsequent damage, because science is bigger than geography and governments. From a scientific standpoint, saving the planet is bigger. It's not their responsibility to save the country."

For some reason that even he didn't completely understand, Jim Kennedy truly believed that saving the country was *his* responsibility.

22

Maneuvers

In 2011, Utah's long-serving Republican senator Orrin Hatch faced the fight of his career, as a tsunami of anti-Washington sentiment crushed Republican institutions. The newly formed Tea Party ran ideologues against incumbents, beating out established politicians in races across the country. Utah senator Robert F. Bennett was the first to go in 2010, after a relentless attack by the advocacy group FreedomWorks that claimed his conservatism wasn't conservative enough.[1] Hatch was next. Activists trashed his record: health care for poor children—*terrible*; a path to citizenship for U.S. military recruits and community college grads—*totally unforgivable*. He'd even voted to bail out Wall Street—*an abomination*.[2] The situation was dire.

"To get through a tough Tea Party challenge, Senator Hatch brought on a new management team," says J. J. Brown, one of Hatch's top advisers at the time. "The new team considered the old guard problematic and began working hard to disempower them." Brown, an eighteen-year veteran in Hatch's Washington office, had successfully ushered dozens of bills into law. He always worked across the aisle, had Democrats as allies, and was a smooth negotiator willing to compromise to get things done. In short, he was everything the Tea Party hated. Facing a future with serious loss of influence on Capitol Hill and the freedom to make legislation happen, Brown worked out a mutually agreeable exit strategy. And just like that, the Mormon father of seven (plus five stepchildren) was out of work. Fortunately, in Washington there are no shortages of jobs for the well-connected. But other than needing a break from Capitol Hill, Brown really didn't have a clear idea what to do next.

It was during that time that Brown got a call from Jim Kennedy and John Kutsch. He remembered them clearly from his time working on thorium regulatory legislation. Hatch's thorium bill was eventually dropped for more pressing issues, but Brown still believed it was an excellent idea. Jim and John brought Brown up to speed on the National Rare Earth Cooperative Act. The topic piqued Brown's curiosity. A lot had changed since he first met Jim and John. Rare earth elements were no longer an obscure topic on Capitol Hill. The Chinese fishing boat incident, the WTO case against China, and Molycorp's promise to break the Chinese monopoly made the metals a sexy topic. Rare earths even made their way into the storyline of *House of Cards*, a Netflix favorite among Washington staffers. "By now, I started to get what Jim was saying, because what he was saying was true. So it really, *really* resonated," says Brown. "I listened to what their issues were—what they wanted to accomplish and what they'd already accomplished." Jim explained that they'd had the interest of Senator Roy Blunt of Missouri, but then Blunt had backed off. Brown was seriously impressed by the traction that they had.

"I'm estimating that out of every four hundred or so good ideas that come to Capitol Hill every year, one will make it into a bill. It takes a huge amount of energy—and usually being right—to actually get your idea into a bill. So Jim and John having a good idea that makes sense and getting a bill is hitting a bull's-eye." What was even more impressive to Brown was that these two unslick Midwesterners did it all without the help of a lobbyist or big money behind them.

"They had really good knowledge of their issue," Brown says. "John obviously knew the thorium energy side. Jim had an understanding of what's happening with rare earths and their relation to thorium." Unfortunately, Brown was no longer in a position to help get Senator Hatch to take up the legislation or support it. "I said, 'Sorry. I don't work for Hatch anymore. I can't help you.'" They said, 'Well, can we hire you to help us?'"

Brown really hadn't thought of becoming a lobbyist before that moment, but it made sense. He had all the skills of a top staffer, so he knew exactly what staffers needed to hear when approached with proposals. He could try to influence Capitol Hill from the outside. He hung up his shingle and called his one-man lobbying firm ThinkPolicy Consulting LLC. Then he signed his first clients—Jim Kennedy and John Kutsch—and went to work on his first piece of legislation as a lobbyist: the National Rare Earth Cooperative Act.

First order of business: Jim's collection of techno-babble documents needed a marketing makeover: "I reformulated the arguments in a format that could lead staffers to the right conclusions. To me it was a really easy thing to do. The topic is interesting, super important, and there is a reason for policy makers to make policy," says Brown. "Sometimes you have to fake that stuff, but in this case, it's obvious."

Second order of business; Jim Kennedy needed charm school. "Jim has little patience for people who don't see what he sees," says Brown. "On Capitol Hill, you're so used to people disagreeing with you that you don't take it personally. You learn to work with people who don't see things the same way you do. Jim's never been forced to do that. He's his 'own businessman' type of guy." Brown says that Jim's proclivity to "tell it like it is" was less than helpful at times. "In the policy world of D.C., sometimes it's a skill to *not* tell it like it is, because you never know when you might need that person later." In Senator Blunt's office, Brown says, "Staffers were like, 'Ugh, here comes Jim. He's right about the issue. But he's such a handful.'"

Fortunately, with his stellar reputation in upper policy circles, J. J. Brown was the perfect person to do damage control. He went to Blunt's legislative assistant, a friend named Downey Magallanes, who worked on energy policies. She'd already given up on the bill. The death knell was the lawsuit between Jim and Alberici. "You don't want your boss picking sides on a nasty legal battle in their home state," Brown says, adding that he would have recommended the same thing to his boss if he were in Magallanes's shoes. She also thought Jim was too much to handle—what Brown calls "Jim just being Jim—impatient and stuff like that."

With the lawsuit settled, Brown thought he could convince Magallanes that the time was right to introduce the legislation. Brown explained the bill's importance in staffer lingo. "Once it's laid out *by* a staffer *for* a staffer, the issue is real. You can't ignore it. When she heard me, she goes, 'Oh my gosh, J. J., I think I have to do this bill again.'"

Rather than introduce the bill on its own—what's referred to as standalone legislation—they would put it into a Department of Commerce bill that included a lot of different manufacturing legislation. Brown found bipartisan support in Democrat Sherrod Brown (no relation to J. J.). Senator Brown's legislative director was very responsive, Brown says, and he had a reputation for getting bills through. "I was very excited," he says. Without having put in a lot of work, it looked like he had a slam-dunk.

Then Washington, D.C., happened. The Commerce bill got bogged down. Senator Brown's staffer left for another job. Even worse, Brown learned that someone from Molycorp got to Senator Blunt through a personal friend of Blunt's. The friend tried to convince Blunt that there was no need for a rare earth cooperative because Molycorp had the whole situation under control. "So that put Blunt on his heels," says Brown. "Senator Blunt had no personal stake in it yet. No mental or emotional stake in it." Brown never had the chance to meet face to face with the senator to present his case. He took it all in stride. It was business as usual on the Hill, and he had two decades of experience seeing even the best legislation derailed by well-connected friends and well-timed phone calls.

Jim, on the other hand, was beside himself. While Molycorp still had the confidence of lawmakers, behind the scenes, the house of cards was falling. Anyone wanting to do due diligence would have easily discovered that Molycorp could never break the Chinese monopoly, because Molycorp had become *part* of the Chinese monopoly. With no place to process metals or make magnets in the United States or any other allied country, the only thing Molycorp could do with the little they were able to produce from previously mined rock, without bringing the mine back online, was sell insignificant quantities of light rare earths to China. The Pentagon's dependence on China for heavy rare earths, metals, alloys, and magnets had not changed at all.

While Molycorp heralded their 2012 buyout of Neo Material Technologies as fulfilling a secure, non-Chinese "mines to metals" value stream, this was a complete ruse. Neo Material was a Chinese company. It had been since 1995, when it was bought from General Motors as the company known as Magnequench. While there were public sources for all these facts—including Molycorp executive vice president Doug Jackson touting China's importance to Molycorp at industry meetings—the company's PR machine continued promoting Molycorp as the end-all, be-all to breaking the Chinese monopoly.

Jim was convinced that there was long-standing collaboration between Molycorp, Neo Material, and the Chinese government. As long as Molycorp was in play, Congress could stand by its free-market principles and take no action to solve the real problem. Brown, on the other hand, believed that Molycorp had stumbled into China by accident. "I don't think they really understood how interested China is in maintaining their monopoly," he says. Brown believes that not long after Molycorp's big IPO, the company realized they'd taken on the impossible. Why was Molycorp still promoting itself as the American solution that would break the Chinese monopoly? Brown isn't sure, but he believes that Molycorp "was still using an old script when the old script wasn't true anymore." Brown says he understands that the job of the company's public relations machine is to protect the company, "but policy makers need to know the truth. It matters to the country." Many legislative assistants that he spoke with expressed deep frustration that what Molycorp promised in 2010 never materialized at all.

In 2014, after realizing that Congress wouldn't pass a Department of Commerce bill any time soon, Brown decided to try for a Department of Energy bill. He soon found that the chair of the Senate Energy Committee, Senator Lisa Murkowski, had a rare earth mine in her home state. Like Molycorp, UCore Rare Metals Inc. was promoting their rare earth mine as if it would break the Chinese monopoly. Apparently Murkowski believed her constituents, so she put a rare earth bill forward that asked for "further study" of the problem and some regulation adjustments. The bill did nothing to establish a domestic value chain. Brown and Jim knew the Alaskan mine would be stuck with China as their only customer too, but that its owners convinced Murkowski to shun the cooperative idea because they didn't want any competition.

Again, normal Washington, D.C., headaches for Brown, who immediately went about calculating his next move. Soon a staffer in Nevada senator Harry Reid's office gave Brown a brilliant idea: the National Rare Earth Cooperative Act should go into the Pentagon's annual policy bill, the National Defense Authorization Act. It was a perfect fit. It would be very easy to convince military staffers that rare earths are critical to defense systems, Brown says, because they are. Plus, unlike the Commerce and Energy bills, the Defense bill gets passed every year.

In June 2014 Brown organized a briefing for staffers—specifically military legislative assistants whose bosses sat on the House and Senate Armed Service Committees. In a large briefing room on the ground floor of the U.S. Capitol, Jim Kennedy presented a slide show to several dozen of the most influential staffers in Washington, D.C. In his presentation were seventy different U.S. military weapons systems. "All of these things can't be made without Chinese parts," he said as he flipped through the slides. "Exoatmospheric Kill Vehicle by Raytheon . . . GMD missile by Boeing . . . RQ-4 Global Hawk UAV by Northrop Grumman . . . MQ-1 Predator by General Atomics . . . F-16 by Lockheed Martin . . ." The normally reserved audience members were on the edge of their seats.

Jim detailed the bill that would give the Pentagon access to rare earth components from trusted sources. John Kutsch outlined how the bill established a U.S.-based development platform for molten salt reactor technology utilizing the thorium as safe nuclear fuel. A Q&A followed. Almost no one asked Jim about how the strategy affected national security (these were all military staffers). The thing that appeared to worry them the most—repeated by several audience members with collective head nodding by the others—was how Jim's plan would affect the "market." Jim assured them it was 100 percent industry-led—from the mines to the manufacturers. Afterward, Brown took Jim and John for meeting after meeting down echoing hallways to Senate and congressional offices. By then, Brown says, "We'd met with probably one hundred military and energy staffers in both parties. Except for one or two, they all responded positively. They all got it."

Everything looked good to go. Brown went back to Senator Blunt's office and convinced them to reintroduce the bill into the National Defense Authorization Act during the Committee markup. Days later, Brown got an e-mail he never expected. It was from Blunt's military staffer, Bo Porsch. The bill was dead. "He told us that the Department of Defense said they don't want it," says Brown. There was no further explanation. (Soon after, Porsch left Blunt's office and took a job with a D.C. lobbying firm that worked for Lynas, the Australian rare earth company.)

Brown had no idea why the Pentagon would not want the bill. For over a year, he'd been trying to get a meeting with someone at the Department of Defense, to no avail. Jim and John suspected that one or more U.S. defense companies opposed the bill due to their dependence on China for strategic parts. They were also banking on big Chinese government contracts to sell defense systems to the People's Liberation Army. In 2011 China had made an example of an American company who dared to rock the boat. When John Kutsch asked the CEO of Danfoss to testify in their briefing, the CEO balked. "They told me that after they testified in 2011, they could not get rare earths or magnets for two years," says John. "It nearly wiped out a big chunk of their business." A company attorney told John they would never testify again. Brown says that he was able to get a long list of support from mining and manufacturing companies, but none of them would go public for fear of the same type of retribution from China.

Brown, Jim, and John went after the Pentagon relentlessly to find out what had happened, but all requests to speak with someone were ignored or rebuffed.

With things flailing on the Senate side after the Pentagon derailed Blunt's bill, Brown found a Texas congressman named Steve Stockman willing to put it forth as a stand-alone bill in the House. "We sat down with Stockman's guy and made some changes to the Senate bill," Jim says. At the last minute, Jim had an idea. He asked Stockman's staffer to add some new language at the end:

> [The National Rare Earth Cooperative Act of 2014] Mandates, beginning in January 2020, all purchased or procured weapon systems to contain only U.S. or North Atlantic Treaty Organization (NATO) member nation produced and sourced rare earth materials, metals, magnets, parts and components. Prohibits the inclusion of any rare earth materials that originate or pass through a non-NATO member nation. Bars any waivers from being granted unless the lead contractor can demonstrate that it has pursued all possible corrective actions, including direct investment into the supply chain.[3]

The bottom line was that the clause would force U.S. defense contractors to buy from "secure"—non-Chinese—sources by 2020. "On our conference call, J. J. warned John and me that the change would piss off the DoD," says Jim. "There was a moment of silence as J. J.'s comment sank in. I could almost feel John and J. J. grinning at the other end of the phone. We unanimously agreed to keep it in."

On June 17, 2014, Stockman's military aide dropped H.R. 4883 into the "hopper" on the House floor.[4] That same day, the bill went to the House Committee on Armed Services for review.

"Almost immediately, we got an invitation for a meeting with the Pentagon's assistant secretary of defense for research and engineering," says Jim. "J. J. and I knew the only reason we got that meeting was because of that clause."

The meeting was a huge score. Assistant Secretary Alan Shaffer was responsible for development of all technology in the Department of Defense, and he had a budget of $11.5 billion. Jim was thrilled. Just three months before, Shaffer had given an impassioned and detailed presentation to the House Armed Services Committee about the state of affairs in research and engineering and begged lawmakers not to cut his budget.[5] "We are the most technologically advanced military in the world," he said, and quoted Secretary of Defense Chuck Hagel: "We must maintain our technological edge over potential adversaries." Shaffer, an air force veteran with a background in math and meteorology, talked of new programs in advanced lasers and guided missiles that the DoD had in development. He gave a rundown of current military systems capabilities. Then he spent a great deal of time talking about the biggest threat to American military superiority: China's advanced weapons systems.

In 2015 China was set to launch 100 new satellites, vastly increasing their navigation and communication capabilities. Long-range missiles in the works could strike as far as Guam. "These ballistic missiles are coupled with advanced cruise missiles that could threaten any surface warfare fleet by 2020," Shaffer warned the committee. In the next decade, Shaffer testified, China's People's Liberation Army planned to have a guided-missile attack submarine that could hit land targets from the sea. Shaffer said that his agency was trying to keep competitive, but he warned that new technologies took years, even decades, and tremendous funds to develop.

Jim couldn't wait for the meeting. He believed Shaffer was the perfect person to understand the severity of China's monopoly. Nearly every defense system he'd spoken about to Congress contained rare earths. On top of it, Shaffer and the Pentagon were clearly up to speed on the growing military threat from the People's Republic of China. Shaffer, Hagel, and Frank Kendall, the DoD's head of acquisitions, would certainly want a secure, non-Chinese source for rare earths.

Once again, Jim Kennedy misjudged his own government.

"When we arrived at the Pentagon, Shaffer was clearly well briefed on the issue," Jim says. But rather than wanting to have a dialogue about solutions, Jim says that Shaffer tried to shut down any serious discussion. About the bill, he said, "We don't know what this is, and *no*, we don't want it.'" When pressed, Brown says, Shaffer told them the DoD was trying to shrink their budget and didn't want anything new on their plate.

"But you *have* to have rare earth components for your defense systems," Jim objected.

"He told us the DoD's official position is, 'We don't need rare earths for our defense systems,'" says Jim.

Jim was stunned. How could Shaffer say that they didn't need rare earths in defense systems? To Brown, it was the same bizarre narrative that he'd recently heard on Capitol Hill. Several high-ranking staffers told him that top DoD bosses came for secret meetings with the chairs of the House and Senate Armed Services Committees and said the same thing. *Everything is under control because we don't need rare earths anyway.*

Jim had his suspicions about why the Pentagon would take such an outlandish position: "They might be hiding the fact that they'd failed to prevent critical supply lines from falling into China's hands," he says. To Jim, in the final analysis, it comes down to two possibilities: a policy failure or a classic cover-up. Brown agrees. "Here on the Hill, people say that that they know the problem is not fixed. Staffers think that the Pentagon's take is ridiculous, and that there's no agency that looks like it's been caught with its pants down more than the Pentagon."

Eventually, Brown says that Shaffer backed off somewhat. "He told us that if we took DoD off as the lead agency, they would be okay with the

Rare Earth Cooperative Act going into the defense bill." But there was one condition: the provision that forbade the Pentagon from buying Chinese parts after 2020 had to go.

"Why do you need us to get rid of that provision if you've really got this problem handled?" Brown asked. He got no answer.

They were getting nowhere, so Jim shifted gears. China was rapidly developing a proven U.S. nuclear energy technology utilizing thorium in a molten salt reactor developed at the Oak Ridge National Laboratory. China would control the commercial distribution of emission-free electricity on a global basis. But this topic was another nonstarter. (In July 2015 Shaffer left the Pentagon and took a job in NATO's Science and Technology Organization.)

Throughout 2014, the Department of Defense continued to claim there was no rare earth problem at all. They had plenty of help perpetuating this myth from policy shops and the media. "New supplies for most rare-earths are coming online," wrote the *Wall Street Journal*'s Joseph Sternberg on August 3, 2014, for an op-ed titled "How the Great Rare-Earth Metals Crisis Vanished."[6] "People like me ran around saying it would all blow over soon enough," Sternberg bragged. Army lieutenant colonel William R. Glaser wrote in a 2014 Naval War College publication that market forces had broken the monopoly, and "U.S. lawmakers need to resist pressures to pass needless legislation that will only disrupt natural market forces."[7] In October 2014 Eugene Gholz, an economist writing for the Council on Foreign Relations, wrote that the "largely successful market response" had solved the crisis, and that "the Molycorp operations in the United States and Lynas operations in Australia and Malaysia have broken the Chinese monopoly."[8] In March 2015 *Forbes* contributor on economics, finance, and public policy Tim Worstall, who calls himself "a global expert on the metal scandium—one of the rare earths" wrote that the "Chinese rare earth monopoly isn't a problem because the monopoly has been broken."[9]

They were all dead wrong.

On June 26, 2015, Molycorp filed for bankruptcy protection. They'd had no operating profit in three years. They'd borrowed another half billion dollars, blown through that, and accumulated astronomical debt. During their short run, insiders reported that Mountain Pass spilled more than 52,000 gallons of chemicals, petroleum, and sewage into the desert. The mine never came online. What little they were able to produce from old surface waste ended up in China.

"Between Molycorp and Lynas, six billion dollars went up in smoke and six years have gone by, and we're exactly where we were," says Jim. Things were playing out exactly the same in Japan. Lynas stock was at $0.04, and the company sent most of their oxides to China. A senior government geologist said that Japanese officials, who'd put $200 million into Lynas, hoping for a secure source of rare earths, felt like they'd been "punched in the head."

"The net result is that Molycorp—and China through Neo Material—used our own stock markets to pump up Molycorp," says Jim. The one rare earth mine in America became another pawn for the People's Republic of China and then disappeared into the history books. "How about that for 'free-market' ingenuity?" Jim says. "We purchased the rope for our own hanging."

23

Zero Sum Game

Having successfully eliminated the Pentagon's opposition to the rare earth cooperative, an encouraged J. J. Brown marched back up to Capitol Hill. In the fall of 2015, he scored a meeting with Lynn Williams, a top staffer on the House Armed Services Committee and the gatekeeper for all strategic materials issues in the National Defense Authorization Act. However, as Jim and Brown didn't have a lawmaker who would directly submit the legislation into the NDAA, Williams gave the pair marching orders, and a few hints. They needed support from the head of the Pentagon's Office of Manufacturing and Industrial Base Policy, from the Energy, Foreign Affairs, and Defense committees, and a request for inclusion from a few House Armed Services Committee members. Walking out of that meeting, Brown was on top of the world. "I said, 'Jim, that's the best meeting we've had so far. She's telling us how to get this done.'"

To Jim—a relentless optimist who expected Williams to put the legislation in right then and there—it was just one more shifting finish line in what had become an eight-year-long series of financial hits, massive headaches, and occasional bouts of utter desperation. He left St. Louis promising Nina this was his last trip to Washington. "If nothing extremely important happened, I told her I would hang up my hat," he says. Once again, he didn't have a bill.

Brown was used to hearing Jim threaten to give up. He couldn't blame him. Late in 2015, after putting $200,000 of his own money into his and Jim's efforts, John Kutsch promised his wife he'd take only revenue-positive work for the foreseeable future. John missed this trip to see Williams, but he kept up constantly by phone and e-mail. With Jim about to throw in the towel too, Brown had to play cheerleader. "I reminded him that Lynn

Williams, a very senior staffer on the House Armed Services Committee, is actually willing to work with us to fix the problem." It was new ground. They had to move forward.

Having Williams on board was not their only positive turn of events. Molycorp's demise had rattled Washington. Legislators who bought into Mountain Pass as the panacea of the American rare earth industry now realized that it was all a mirage. Molycorp's antagonistic exit strategy made headlines in September 2015, when company executives gave themselves $2.9 million in bonuses in the middle of bankruptcy proceedings.[1] Even better for Jim, after the Pentagon claiming for so many years that the situation was under control, the truth had finally come out. This time the Department of Defense itself was pointing the finger. A 2014 internal investigation by the DoD Office of the Inspector General said that the rare earth debacle was not solved at all—that the agency was woefully unprepared to deal with this grave supply issue.[2] It appeared that in the 2012 report to Congress, Frank Kendall, undersecretary of defense for acquisition, technology, and logistics, had signed off on the Pentagon's rare earth readiness based on bad data and false projections propagated by Molycorp—a single company that was now completely bankrupt. It was this report that had shot down Jim's first attempt at getting the legislation passed with Senator Blunt as the lead. In Jim's words, "Like so many investors, the Pentagon was burned by what amounted to a Wall Street pump and dump con job."

"The Inspector General report confirmed what Jim was saying all along," says Brown, who was not aware of the classified report until a year after it was quietly made public. It felt great. Brown believed that the internal investigation—called for by someone on the House Armed Services Committee—likely stemmed from the work that he, Jim, and John were doing. "We've talked to just about every relevant staffer from both parties and all the important committees," he says. With the DoD calling out its own failure and no viable non-Chinese rare earth source anywhere in the world, Brown felt very good about the future of the National Rare Earth Cooperative Act. In a round of meetings during the fall of 2015, Brown and Jim saw much more support, even inside the Pentagon. "With Molycorp gone, everyone was listening because no one had a backup plan," Brown says.

Brown continued his assault on Capitol Hill, getting his ducks in a row on both the House and Senate side. He sought out lawmakers who would back or introduce the legislation, and influential insiders who could help persuade lawmakers it was vitally important to do so. Brown and Jim were in overdrive when Brown landed an especially big fish to help their lobbying efforts—a retired army general by the name of John Adams.

A quarter century had passed since Major John Adams left the English department at West Point to take his post as a targeting officer in the first Gulf War. After twenty years deployed mostly abroad, Brigadier General Adams retired in 2007 to a completely different America. Hoping to use his expertise and connections to help Washington make better national security decisions, he set up his own consulting shop, Guardian Six. One of his first big clients was a nonprofit, nonpartisan advocacy organization called the Alliance for American Manufacturing. The group's constituents were some of America's leading manufacturers (among others, U.S. Steel and Huntington Ingalls Industries) and the United Steelworkers union, who'd been fighting a losing battle against the tide of offshoring since the 1990s. Adams decided the best way to help was to take a look at American military products made in countries that could manipulate the supply chain. He hoped his work would motivate Congress to bring back manufacturing jobs to the United States and demand that the Pentagon buy from reliable domestic sources.

Adams—who spent most of his career abroad and with his radar off procurement issues at the Pentagon—couldn't believe what he discovered. "It was shocking that we had so many vulnerabilities in so many important technologies," he says. "We'd outsourced so many critical defense items: semiconductors, advanced magnets, biological devices." There was even a type of rocket propellant that the Pentagon couldn't get anywhere outside of China.

Two of the fourteen major "nodes" Adams chose to investigate for his report were rare earth metals and advanced magnets.[3] Both presented grave vulnerabilities that no one was paying attention to. "One of the things globalization's done," Adams says, "is cause a real lack of effective management of our defense supply chain. If the supply chain runs to Canada or another ally, then perhaps that's not of very great concern. If they're running to strategic competitors—that's a matter of great concern to me as a military officer."

When he went to Washington to sound alarms, Adams was disappointed by what he found. Many Pentagon bosses and lawmakers seemed too overwhelmed to act. Some even appeared too apathetic to care. "Rare earths are a great example of, first of all, denying that there's a problem, and second, being unwilling to do anything about it because it's too hard," says Adams. Unlike the response Jim Kennedy got from the Pentagon, the general's welcome was decidedly more cordial. "The overwhelming reaction was 'Tell me more,'" says Adams. "They were doing everything they could do to wrap their minds and considerable analytical skills around the problem." Adams believes that a lack of manpower contributes to the impotence. "We're talking maybe two dozen in the Pentagon's Office of Manufacturing and Industrial Base Policy that are charged with keeping our defense industrial base healthy. I did get a sense that there are a lot of issues that are too hard to address."

When Adams approached defense companies with his concerns, the situation seemed even more dismal. "I think that defense contractor CEOs are most concerned with keeping prices down and competing for contracts," Adams says. But were they concerned about China's dominance in so many critical defense components? "Truth is, they don't really care. They want to get things at the lowest cost so they can sell at the lowest cost. I get that. But we have to look at what's good for the United States." Adams says that when Pentagon senior leadership positions are filled right out of the defense industry, it creates an "incestuous" revolving door. While Pentagon leaders should be minding the store, strong ties to the defense industry arguably make for cloudy judgment. "I do believe in a global economy, but it's got to work for us. If it's not working for us, it's working against us. It's a zero sum game."

Like many aspects of twenty-first-century American politics, the problem was economic. "We're in a trade war with China, and we're the only ones that don't know it or don't act like it. We're just like, 'Okay, well . . . we don't want to put any kind of restriction on the free market.'" Still, Adams found very few who didn't admit the issue was grave. "Whether we're talking about Congress or the Department of Defense, there seems to be a general agreement that there is a problem."

Why don't they do anything to fix the problem? China is an all-powerful Daddy Warbucks to the defense industry, selling parts and labor cheap and heaping riches on defense CEOs and their shareholders. But China can pull the strings at a moment's notice and shut down everything, from lasers to missiles to satellites. Adams isn't a big government guy, he says, "But some industries are so important to our national security, losing control of them would put us in a hostage position. I'm a soldier, so I look at the world strategically.

"I have no doubt that, at some point, China will realize its desire to be the strongest global economy," Adams says. "And just as they do now with our debt, they will be able to coerce us and say, 'So you really don't like that we're going to invade Taiwan? What are you going to do about it?' That's where we're headed." America's best way to avoid a conflict with China is simple, Adams says. "Don't let them dominate our economy and our defense industrial base."

By the time Adams got a phone call from J. J. Brown, the general was well aware of the deep fissures cutting into America's national security. What Adams didn't have were answers on how to suture them back together. So when he heard the details of Jim's rare earth cooperative idea, Adams jumped on board to help. Adams believed—like Jim Kennedy—that there was no way that the U.S. military could survive if it had to play by the economic rules of Wall Street.

Brown was thrilled to have the general on board; Jim even more so. After so many years with just John by his side, the rebels from the Midwest would be taking meetings with a retired general carrying a 400-page playbook on the massive failures of American national security. This was sure to up their game and help with their current efforts to get support from defense contractors. Many CEOs liked the cooperative idea, but Brown still couldn't find anyone to go on record for fear of retaliation from China. It was what Brown called their Achilles' heel. Brown hoped that a decorated military veteran with an impressive CV would be enough to override the corporate silence.

In early 2016 Jim Kennedy geared up for his final assault on Washington. He was eight years in and hundreds of thousands of dollars out. Two years past fifty, he still had his boyish good looks and only a few grey hairs. But he was exhausted. The fight had taken a toll. The more he learned along his journey, the more he realized how much was lost. The harder he slogged up the mountain, the more he understood how far his country was slipping behind. The missteps and failures list grew so long, it boggled his mind. There was a nuclear renaissance and a clean energy promise that never came to pass. America's rich geology ushered in an unbridled industrial revolution in the early twentieth century. But by the 1970s, after Nixon ended the gold standard, American iron, copper, and rare earths tried desperately to compete with global markets. Most failed. Jim still felt sick when he remembered how antiscience fervor captured Washington in the 1990s and left once-thriving national laboratories more museums than research institutions. For thirty years, politicians had been taking a knife to the belly of America's great manufacturing cities with trade policies that turned political donors into billionaires. The stock market that helped build American businesses since the early 1800s was now a reckless, limitless Ponzi scheme. China put out press releases and reports outlining their goal to dominate the world. Washington yawned. When it happened, those at the very top of the American government acted like they never saw it coming. Kids were sent to wars they couldn't understand, and the Pentagon couldn't protect them unless they bought parts made in China. America today was a severed limb of the America that once was.

Jim sometimes wished he didn't, but he cared about all of it. He cared a lot. He sometimes wished he could stop the torture of his churning brain, but he couldn't turn off his thoughts. Sometimes he felt very alone. "Unfortunately, our political and media culture refuse to evaluate anything deeply, or in a historical context," he says. "Everything is measured by money—the proxy and arbiter of all things." Many times, the struggle seemed pointless. "Regardless of the insanity, I keep working every day. Being Irish, I don't have the sense to give up when I've lost. I refuse to surrender."

While every setback during his eight-year-long endurance race took a few years off Jim Kennedy's life, Brown believed that things were moving along quite nicely by Washington, D.C., standards. "My role is to figure out what's possible, what can be done and what levers need to be pulled and pushed to get us there," he says. "But a lot of the energy and brilliance and resources comes from Jim and John. No one would be talking about this issue if it were not for them." Brown ticks off a list of the pair's accomplishments: "Jim was invited to speak at the International Atomic Energy Commission; John speaks at nuclear energy conferences across the United States and at national labs. Jim's a key speaker at the world's largest mining conference year after year. Nobody else is out there doing this the same way. Jim and John are the only people putting on briefings and getting in people's faces."

Brown says that while Jim developed a new political correctness in meetings on the Hill, he still went off behind the scenes. "But I never blame him for that," says Brown, who believes that Jim's motivations are far from selfish. "To be honest, I think he's doing it to save the country. I think he'd have a hard time admitting it, but he's a plain old patriot, which makes him all the more angry when lazy staffers don't pick up his issue and see how important it is. If we ever solve this problem, there really ought to be a monument to him, because nobody else is out there pushing a solution."

In February 2016 Jim arrived at Reagan National Airport. Like he had so many times before, he took a cab to the Rayburn House Office Building on the south side of Capitol Hill. John was by his side. He couldn't resist coming to help, so he'd booked some work in Washington for Terrestrial Energy, a Canadian company hoping to build the West's first commercial thorium molten salt reactor. Like Jim, John couldn't give up, however frustrating and time-sucking it all was. "You don't generally find this mindset in the United States anymore," he says. "The idea that we can be great and get things done and build stuff and be the best in the world. I believe we can. I'm still stupid that way."

If there are real-life superheroes, they most likely do not have flowing capes and muscles bulging through glistening spandex. Instead they probably look more like the four men walking down marble hallways that winter day in 2016: Brown, a conservative Mormon and father of twelve children who spent his adult life methodically working an increasingly incomprehensible system for the betterment of his country; Adams, a Desert Storm war veteran who felt betrayed by American politicians dragging kids into wars, only to leave the Pentagon precariously unprepared to protect them; Kutsch, a loud and animated engineer with a serious affinity for creating and inventing, and desperate to see his country do the same; and Jim, a rebel intellectual in cowboy boots with the power to see what's wrong with the world and a well-calculated idea about how to make it right.

This time felt so different. Jim Kennedy was no longer Chicken Little, begging lawmakers to notice the sky falling; all of his predictions had come true. The sky had dropped on Molycorp, then Lynas, the promised solution for Japan. The DoD's smokescreen disappeared with its own internal investigation. Jim even had a report in his pocket from the Government Accountability Office, which agreed that the DoD had no secure rare earth sources at all. The report estimated it could take fifteen years for the United States to rebuild its own rare earth value chain. Congress had to act.[4]

To Jim it all pointed to his rare earth cooperative as the answer. With staffers on Capitol Hill hearing them like never before, and the general echoing Jim's call for the cooperative, he felt something he hadn't for a long time: hope. They did have one concern on the House side: Lynn Williams, the staffer who Jim says "got it" and was paving the way for the legislation, had left her position. Still, they had more than two dozen excellent meetings and tons of support. Since they couldn't convince any one lawmaker to personally push for the legislation's inclusion in the NDAA, they needed serious support from multiple fronts.

Getting in the House defense bill would give them an excellent chance of making it through to final ratification by President Obama in the fall. The final bill would be a combination of both the House and Senate versions. So to cover all their bases, they went to lobby the Senate side as well.

Things didn't go quite so smoothly. "Sure," said a minority staffer on the Senate House Armed Services committee, "the situation needs to be fixed. But this is someone else's responsibility. Try the administration. Try the Energy Committee . . . maybe the Commerce Committee, anyone but the Armed Services Committee." The other staffers backed him up.

Jim's face grew red. He openly scowled. His eyes narrowed. J. J. Brown had ducked out early for another briefing. There was no one to stop Jim. In the most contemptuous tone he could, Jim began:

> We are private citizens who are bringing you, the members of the Senate Armed Services Committee, a well-documented national security issue. We have already been to the administration with this. They did not act. Instead they pushed the World Trade Organization case, which only made things worse. We have been to the Pentagon. They will not even recognize there is a problem. In fact, according to the former assistant secretary of defense, Al Shaffer, the Pentagon's official position is that rare earths are not critical to U.S. weapons systems—completely inaccurate, and we all know it.

John Kutsch held his breath. Jim wasn't finished. "Even the government's top auditing arm, the Government Accountability Office, has classified this issue as a 'bedrock national security issue.'" He paused for effect and made eye contact with each staffer. The staffers glared back. Jim went on, "We are here, now, because it is *your* responsibility to fix this. This is not *my* responsibility. This is not *John's* responsibility. *It's yours.*"

Leaving the meeting, Jim had a lump in his throat. He didn't regret speaking the truth. What made him feel bad was that he hadn't kept his cool with General Adams in the room. He had so much respect for the general, and suddenly he was back in the army, a punk recruit disappointing a senior who he greatly admired. Jim apologized immediately once they were outside. "I told Jim that he did just fine," says Adams. "But that I probably wouldn't have expressed my frustration is such a manner." Still, the outburst did nothing to diminish Adams's respect for Jim. "I would never fault him for his candor and passion. Jim is a great American, and I am proud to know him," he says.

Brown found out later about Jim's outburst, but was long past admonishing Jim, possibly because Jim said everything that Brown was feeling but a refined Washington lobbyist couldn't say, in language that a polite Mormon dad wouldn't use. They were clearly not going to get support from the Senate Armed Services Committee, but Brown was still encouraged about their chances in the House bill. He crossed all the t's and dotted all the i's he could think of. Then all four superheroes took a collective deep breath and hoped for a miracle.

There was nothing to do but wait. Back home in St. Louis, Jim went upstairs to his home office over the garage and sat for hours at his computer, searching the news. He sent out e-mails to John and Brown with the articles he found, accompanied by his "Jim Kennedy" commentary. An article saying that the Pentagon was still denying a crisis: "This demonstrates a strong institutional repulsion to reality-based geopolitical thinking in asymmetric warfare." A piece on Japan looking at drilling into the sea floor for rare earths: "Sheer stupidity." A Heritage Foundation report that "the market is working" to solve the rare earth crisis: "The Heritage Foundation is the mouthpiece for failed Friedman economics. They have nothing useful to contribute to reality." A news item on Russia and China collaborating on rare earths: "Will China/Russia Crush U.S.?"

Jim woke up early every day and drove to the Bourbon Lodge. He hung out with Emilio (the caretaker of fifteen years and a much-loved member of the Kennedy family) and worked on his list of long-neglected maintenance issues. In the afternoon, he drove Dravin an hour each way to her horse barn. He chauffeured Shale to various sleepovers and hangouts. As always, he dodged cultural events and fundraisers with Nina's social circle when possible. He raged against the machine with his once politically conservative father-in-law, Nabil, who was still one of his biggest supporters. He spent time with his dogs. He gardened. He smoked a cigar every night in his backyard. When his three girls left him alone in their sprawling St. Louis home, he ate popcorn for dinner and watched political documentaries and action movies.

In late April 2016 the House released their version of the National Defense Authorization Act.[5] It was little covered in the media. Most Americans probably didn't have a clue about the hundreds of line items and amendments that Congress wanted $610 billion to fund. There was an amendment barring the military from doing business with institutions that fly a Confederate flag. One mandated that there be a "hearing" on the security risk of foreign-bought parts in Pentagon satellites. (According to Jim, a cover-your-ass bill introduced by lawmakers who refused to do anything concrete but wanted to look as if they cared.) There was even an amendment to force the Pentagon to buy American-made sneakers for all new recruits—a boon for two Maine senators and their constituents, the New Balance footwear company.

Jim scrolled quickly through the list. Then did a search for the term *rare earth*. Then he sat back and felt his body grow hot. It was almost too much to fathom. While the Pentagon would be legally required to source athletic shoes made in America, for the rare earth metals, magnets, and lasers essential to every single high-tech system in the U.S. military, there would be no such law. Incredibly, there wasn't a single mention of rare earth elements in the bill at all. For the biggest defense companies in the world supplying the most powerful military force on the planet, the case of rare earth elements would be business as usual—brought to you by the People's Republic of China. And for one brief moment—perhaps a microsecond—Jim Kennedy was ready to quit.

Then he went into overdrive. With Brown, Kutsch, and General Adams by his side, Jim changed tack. Forget the worthless, do-nothing Congress; he would take the advice of the Senate Armed Services staffers he'd sparred with. The rare earth cooperative didn't need to be written into law by Congress. It could come to life with the stroke of a pen: *a Presidential Executive Order*. In August 2016, Brown scheduled meetings with the National Security Council and the Office of Science and Technology Policy. Jim handed off a 400-page dossier he'd put together in a matter of weeks detailing the plan and its implementation. General Adams impressed them with the strategy's vital importance to national security. John outlined how the cooperative would help usher in a clean energy revolution. Their impassioned pitch was refined and bolstered by the honesty of their mission. But the wind was no longer in the sails of the Obama White House, and the positive feedback they received in the Executive Office Building faded away on the winds of a waning administration.

Then on November 8, 2016, Washington, D.C., turned upside down. A bombastic real estate mogul and television personality rode a populist wave to become America's 45th president. The centerpiece of Donald Trump's campaign consisted of trash-talking China and disparaging trade deals that screwed Americans out of decent jobs. To many, Trump's January takeover of the Oval Office felt like doomsday cometh. But J. J. Brown couldn't have been happier. Like many of his fellow Mormons, Brown was not a huge fan of the new president. But it appeared that Washington was coming out of a deep Tea-Party-infused coma. Republicans, whose main job since 2010 had been to sit around and make sure that government did nothing at all, were on notice. Things were happening. The air was electric.

D.C. had its groove back. To Brown, who so missed his days passing meaningful legislation as one of Capitol Hill's more prolific staffers, the nation's capital was nothing short of a phoenix rising from the ashes.

Jim, John, and Brown went looking for support for the rare earth cooperative in Trump's transition team. Jim connected with Ned Mamula, a resource expert he'd met at the ultra-conservative Cato Institute. Mamula had been tapped for Trump's transition team in the Department of the Interior, and asked if he could include Jim's proposal in his portfolio. Brown had friends and former colleagues in high places and tons of respect from Republicans who were now running the whole show. He got a buy-in from a guy on the National Security Council. Then things got even better when Trump tapped billionaire business investor Wilbur Ross for Commerce Secretary. Brown always believed that the cooperative belonged with the Commerce Department, and Ross would be a perfect candidate to usher it in. Better still, on board to head Trump's new Office of Trade and Industrial Policy was economist Peter Navarro, a huge critic of unfettered globalization and vocal proponent of the U.S. getting its balls back from China.

In the winter of 2017, Jim and John watched as the play they'd been drafting over the last decade was cast while Brown finalized the orchestration. The stage was perfectly set. Act One would open with Wilber Ross and Peter Navarro delivering an executive order to President Trump. With the stroke of a pen, the very first cooperative among the West, Japan, and South Korea would be created to counter the crippling power amassed by the People's Republic of China. With one well-placed signature, the pieces of progress would begin to coalesce and rebuild the technological prowess of America and its allies. In the final act, Jim Kennedy and John Kutch would begin the project they believed it was their destiny to create. It all looked perfectly ready to go, just like that. And like so many times before, all there was to do now was wait.

Epilogue

On the Shoulders of Giants

February 2017

For the past two years, I've had a front-row seat to a real-life David and Goliath battle. I wrung my hands and clenched my teeth watching Jim Kennedy fight powerful institutions that were driving our country head-on into a hopelessly unprepared future. Sometimes, I felt so much despair with the direction our country is going, I pretended it was all fiction. Still, as emotionally hard as it's been to see our leaders' missteps through Jim Kennedy's eyes, I feel incredibly privileged to have encountered the cast of thoughtful and brilliant people whose lives somehow wound their way into this story.

When I set out to write this book, I was so distant from the epicenter of rare earth elements, I had little idea where to begin and no idea who I might encounter along the way. Then I stumbled upon Jim—and in turn, John Kutsch, Peter Leitner at the Pentagon, and General John Adams, men who didn't swallow the same propaganda that caused others to remain silent as our government sold us out. My worldview expanded as I got to know China's history and people, a culture diametrically different from ours. The Chinese are so rooted in the past that they see the future as something to protect long after their own lifetimes. Their government is also very calculating. Once, at a conference in Virginia, Jim railed against the United States for selling out to China. One guy in the audience said, "Why do you hate China so much?" On the contrary, Jim said: "I'm jealous of China. I wish America would take care of its own interests like the Chinese do." Me too.

Had we taken a longer view of our collective future, things would be different for many who found a bit of their lives documented here. Back in Oak Ridge, Tennessee, Dick Engel, one of the lead molten salt reactor inventors, is no longer surprised by America's lack of forward thinking in nuclear energy—but he's not happy about it. He's grumpy. He personifies the word *curmudgeon*. He says, well, if *we're* not going to develop new nukes, at least *somebody* is. He seemed disappointed that he wasn't invited to Oak Ridge the day Jiang Mianheng came courting his chemist's reactor. His old buddy Uri Gat, who ran the Oak Ridge molten salt program back in the 1960s and '70s, is still full of hope that nuclear energy will save the environment. Engel shrugs his shoulders. The world is way overpopulated, he says. We're all pretty much doomed.

After their synchronous magnet discoveries in 1982, John Croat and Masato Sagawa left their respective companies and set up shop for themselves. The shifting balance of global power left scars on both their careers. When Croat tried to start a new business in 2008, he was sued by a company claiming to own the Magnequench patent. Believe it or not, the company suing Croat—after buying out Chinese-owned Neo Material (formerly Magnequench)—*was Molycorp*. Wrap your head around that. The problem stemmed from what Croat calls the "gross incompetence" of the U.S. Patent Office, which often allows big corporations with teams of lawyers a decade or more past the seventeen-year patent protection allowed by law. He eventually won, but the legal fees completely broke him, and his company never got off the ground. He spent the last years of his career working as a consultant in China and then retired to Naples, Florida.

In 2010 Dr. Sagawa developed a new technology to mass-produce his sintered magnets. He also had a hard time getting a license for his company to produce the magnets—long after the patent should rightfully have expired. Many in the worldwide permanent magnet community feel that Sagawa and Croat deserve the Nobel Prize by now. I have to agree with them. No one ushered in a technological revolution like these two men. But history forgot them both. The only profile I found, googling John Croat and his permanent magnet invention, was from his grammar school newsletter. I'm not sure why. My guess is that in America's rabid determination to make money from new inventions, financial return is celebrated with such fervor that it steals the thunder of those who actually made it possible. I hereby nominate both of these great men for the Nobel Prize in Physics.

Something else has changed since I began to write about science early in my career, and it began long before Trump took office: I now encounter government scientists who are no longer allowed to speak freely to reporters. This was never the case in the past for those not working on secret projects, and it's terribly unsettling. One of my most important sources to fact-check the geology and chemistry in this book was a U.S. Geological Survey geologist. He spoke fairly openly to me for background, though he told me he was supposed to get clearance even for that. If I was going to quote him, he said I would have to pass it by the government media relations people to check if what he said was okay with them (I know from experience that most media relations people have no science background at all, which make this a cumbersome and irritating roadblock). I can tell you that getting an interview with Cheryl Seeger from the Missouri Geological Survey was like getting inside the CIA. When I finally got clearance, I asked Seeger how budget cuts affected projects at the Missouri Bureau of Mines. She looked nervously sideways to a stern PR handler and asked if she was allowed to answer that question.

I believe that Jim's friendship with Cheryl Seeger was a casualty of this disturbing clampdown. She and Jim were thick as thieves for over a decade, but now they rarely talk. When Jim brought two Japanese government scientists to the bureau headquarters in Rolla, Cheryl agreed to meet with them, but Jim had to swear he wouldn't discuss anything political at all.

Everyone I spoke with who worked closely with Jim holds him in high regard. Some say he's a hothead. Some say he never shuts up. Some say he's charming and funny. All say he's super smart. Brad Van Gosen, from the U.S. Geological Survey, wondered if Jim's one-track mind ever deviated from rare earths and thorium. One day over a late dinner in Vienna where the two were speaking at a nuclear conference, Jim finally opened up, and Brad found him to be highly entertaining. Even after Stan Short was put through a harrowing year of subpoenas and harassment by Alberici Constructors' team of attorneys—even after he learned he'd lost a huge deal because of something that Jim and Nina hadn't been able to tell him—he still has a very high opinion of Jim. The two have coffee together every so often.

One thing Short mentioned that I found equally true in my dealings with Jim: Jim will never admit that he was wrong about anything. If something doesn't work like he thought it would, or wasn't the way he said it would be, he seamlessly moves along to idea B, C, D, completely erasing anything that didn't go as originally planned. These days, Short spends more time with his daughter and grandkids than putting together financing deals. But he's always on the lookout for something cool, he says, something amazing like Jim Kennedy's plan to transform Pea Ridge.

In general, Jim's old friends seem amazed by his life trajectory. Rich Todaro—who came to work for Jim and his father as an intern—did so well at Kennedy Capital, he retired in his early forties. The pressure Jim puts on himself to save the world seems foreign to Todaro, who lives in a sprawling home in St. Louis with his wife and two daughters and rebuilds classic cars. Jim's high school buddy Mark Haselhorst hadn't seen Jim in years when Jim called and asked if he would meet me for an interview. *Mark, listen, this woman is writing a book about me. I'm about to leave for Geneva because I'm giving a talk at a United Nations thing after I have this meeting with the White House guy on technology. China has got America by the balls, Mark . . . I have to do something.* "I knew that there was some mental illness in his family," says Mark. "I'm like, 'Holy shit, he's lost his mind. This is so sad. Poor Nina.'" When I met Mark at Jim's house, he was still blown away by Jim's path in life. The Freak was *for real.*

There are, of course, people who don't like Jim at all. Some are even related to him. But they either wouldn't talk to me, or I chose not to interview them. In the case of Jim's siblings, I left the decision to Jim and Nina in the interest of keeping peace in a very fractured family. Some who trashed Jim's credibility in Washington wouldn't talk. One was Jim Sims, the storied Molycorp spokesman who Jim says badmouthed him endlessly and did everything to derail his legislation. I e-mailed Sims in the midst of Molycorp's precipitous slide. Sorry, Sims told me, we're too busy getting our processing plant up to have any journalists come visit. A few weeks later, I saw a fluff piece by some hapless California reporter who toured the mine that would save America from China. Sims is certainly a genius at spin. You can argue the morality of it, but the public relations business is one of the most important investments for billion-dollar corporations these days. *Scandal*'s Olivia Pope is a real deal. How else could we forget that General Motors, America's number-one car manufacturer, and the company that supported John Croat's world-changing invention with an unlimited budget, was responsible for the deaths of 124 Americans because they wouldn't change a part that cost a dollar?[1] In just the same way, Jim Sims spun a pile of worthless sand into a Wall Street sweetheart and Pentagon savior.

One of the most prominent consultants making market predictions and vouching for rare earth mining start-ups is Dudley Kingsnorth. Kingsnorth, more than anyone, is the analyst who inflated the rare earth bubble with breaking forecasts and projections. The more I looked into it, though, the more suspect his reporting seemed. Apparently it did as well to a German fighter pilot turned Ph.D. candidate named Volker Zepf, who dug deep and found that Kingsnorth and others were all citing the same ten-year-old USGS data to make market predictions.[2] Why didn't they have any new numbers? Because China controls the entire value chain, nearly all rare earths go into products manufactured in China, and China has no interest in releasing data so people can make predictions to encourage new rare earth projects. Toward the end, Kingsnorth's margin of error was plus or minus 25 percent, meaning he could be off by *half*. "In summary this situation is, from a scientific point of view, neither reliable nor valid," wrote Zepf for his Universität Augsburg doctoral thesis. "Yet, all reports and scenarios are based on these figures."

If there is a prize for someone who exposed a bunch of flawed statistics that snowed Wall Street, the Pentagon, think tanks, media outlets, and Japanese and U.S. lawmakers, that award goes to Volker Zepf. Congratulations, Volker. I wish it had done some good. But apparently, your thesis is too difficult to read. Forecasts based on wishful thinking are much easier to swallow.

I saw Kingsnorth a second time at an industry meeting in Chengdu, China. A Chinese government representative gave a presentation about China's planned rare earth collaborative, and Kingsnorth asked if Western companies would be allowed to participate. Sure, the speaker said—*as long as you move your business to China.* As Kingsnorth sat back down in the massive, opulent conference room, the speaker's final slide projected this sentence high above Kingsnorth's head: *He who controls the resources, controls the conversation.*

Another difficulty I had was sifting through mountains of flawed think tank and policy shop reports and the proliferation of propaganda that they've spawned. One case hit me personally. After being invited to complete a long proposal to the Alfred P. Sloan Foundation—a large foundation with the goal of enhancing the "public understanding of science"—for their book grant and waiting for months and months to hear back, I was turned down in a two-sentence rejection e-mail. A month later, a paper came under the Council on Foreign Relations by an associate professor of public policy at the University of Texas. Professor Eugene Gholz reported that the "largely successful market response" of Molycorp had saved the day, and China's monopoly was no longer a problem.[3] This was six months before Molycorp's bankruptcy. The company's contribution to breaking the Chinese monopoly was *zilch*. I felt sick to my stomach when I read who'd funded the report: the Alfred P. Sloan Foundation. I wrote a long e-mail to Doron Webber, the Sloan book contact, explaining that Gholz's report was completely flawed. I signed off: "Sadly, this report got a lot of press and continues to perpetuate false narratives about a critical scientific and geopolitical topic. In keeping with the Sloan Foundation's mission to advance the public understanding of science, I would very much like to have the foundation's support to finish working on my book about rare earth elements in a way to truly shed light on a very complicated and important topic." I never heard from the Sloan Foundation again. If they want to give this book an award for actually telling the real scientific and policy story of rare earth elements, I will happily donate any prize money to a fund to help science journalists investigate real science.

It may have been any number of erroneous "expert" analyses that prompted *Wall Street Journal* reporters John Miller and Anjie Zheng to come up with a real doozy of derivative hogwash. The ink wasn't even dry on Molycorp's bankruptcy when Miller and Zheng reported that Molycorp's bankruptcy didn't matter anyway because the rare earth problem was solved.[4] "Battery and magnet makers found alternatives to rare earths." *Wait . . . What? Seriously?* Since this was news to me and everyone I knew in the scientific community, and the authors listed no source at all for this as-yet-unreported miracle of science, I e-mailed them both. Both ignored me. I no longer have any faith in the veracity of mainstream media science or economics reporting, and I expect to see an unsourced article on cold fusion in the *Wall Street Journal* any day now.

Some others I mention in the book refused my interview requests, like Valerie Grasso of the Congressional Research Service, who Jim sent several heated e-mails to. Rick Lowden—the person at the Pentagon who first met with Jim, and later told Jim not to contact him again—"respectively declined" my request. The DoD Inspector General press person answered a couple of my questions with one-line answers, while saying that the other 90 percent were "beyond the scope of this report." Jim gave me the number of a Boeing guy whose job was to secure rare earth parts from China for Boeing and other defense contractors, and who'd told Jim off the record that he couldn't do anything to piss off China. He was speechless when I called. He couldn't comment on anything, he said, and sent me to Boeing's Washington, D.C., person. I left a message for her, which apparently prompted a call from Boeing's media relations person, who had me write up long, detailed questions that Boeing then completely ignored. Through a long, convoluted path, I finally got to former Indiana senator Evan Bayh's assistant, who promised me an interview. Bayh was the one fighting to keep Magnequench in Indiana, but he'd left Congress in 2011 to become a lobbyist and Fox News contributor. The interview never materialized. Former California congressman Christopher Cox never got back to me. I was curious how he felt that his shocking and foreboding 1999 report on China taking over America's technology had basically gone nowhere.

In late 2015 I sat in on a press conference put on by Defense One, an online publication for the defense industry. On the panel was none other than Frank Kendall, Pentagon undersecretary for acquisitions. The audience got cards on which to write questions that would be hand-picked by the moderator. A woman from the British embassy sat next to me, sensed my desperation, and gave me her card to double my chances. After Kendall went on and on about U.S. military superiority, the Q&A began. Somehow the stars were aligned, or perhaps the Defense One moderator began to read my brief question before fully digesting it: "How will the United States maintain military superiority when China has a monopoly on critical materials like rare earth elements?" The entire room went silent. I hadn't even turned on my recorder because I was 100 percent sure that with this audience full of Lockheed, Boeing, GE, Halliburton, and General Dynamics reps, my question would never be asked. These were all the companies that buy parts from China and dance around getting clearances. The presser was live-streamed. Someone tweeted: Defense One WENT THERE! Kendall looked annoyed. "We *have* secure sources for all critical materials," he stated emphatically. "Would you like to elaborate?" the moderator asked. No, Kendall told him.

Most rejections of my interview requests, I half expected. But one broke my heart. When I realized that I would never land an interview with Dr. Jiang Mianheng, and I would have to build his story based on other people's accounts and very few public documents, I wondered if he was the right character to personify China's meteoric twenty-first-century rise. But the book kept pulling him back in. I hope I did his story justice without the romanticization that sometimes filled my head (especially when I found out he's buddies with Elon Musk). I learned in May 2016 that Jiang no longer heads the thorium reactor project for the Chinese Academy of Sciences. I was worried and a little sad. It seemed like, after Alvin Weinberg, it would be Jiang's baby to see to completion. Searching for news, I found a few reports about former president Jiang Zemin (Jiang Mianheng's father) being a target of current president Xi Jinping's corruption crackdown. One report said that Jiang Mianheng was caught up in it all.

Fortunately Chen Kun, who worked under Jiang Mianheng and is still on the thorium reactor project, told me he hadn't seen any such thing. He told me Dr. Jiang left the Chinese Academy of Science at typical retirement age and is busy running the new ShanghaiTech University, a small state-of-the-art campus designed by Jiang, patterned after CalTech, and open to high-level Ph.D. candidates only. Jiang's real passion, says Chen, is educating a new generation of great Chinese scientists.

Chen is still involved with the thorium molten salt reactor and collaborating with researchers at Oak Ridge, Massachusetts Institute of Technology, and other American universities. David Holcomb is still being paid by funds that come from the Chinese Academy of Science, and both Holcomb and Chen agree that it's been a fruitful collaboration. Back home in China, Chen and his team are trying to land a location on the northern coast of China for their first test reactor. The Chinese Academy of Science is going through a big shake-up, Chen says. The thirty-five-year-old researcher thinks the institution is too old-school for the new China. He's waiting to see how things shake out.

With all the talk about China's slowing economy (which is still growing in a positive direction at three times the United States' rate), I asked Chen what things were like day-to-day in Shanghai. He said he definitely feels the slowdown. For the first time in a long while, there are empty stores and shopping malls in Shanghai's opulent districts. Unfortunately for Chen and his wife (both scientists on fairly modest salaries), prices for everything from rent to food are still sky-high. He does not see his government opening up to Western-style social "freedoms" (like Facebook) in the near future, but some things have progressed. For the first time in thirty-five years, in 2015, the central government lifted the ban on having more than one child. Chen and his wife are thinking about adding a baby to their family of three.

Peter Leitner, inside the Pentagon from Reagan to G. W. Bush, is still based in Washington, D.C., and running two think tanks, the Washington Center for Peace and Justice and the Higgins Counterterrorism Research Center. I asked him how he feels about how things are today, nearly a decade after he left the DoD. He sent me an article from a blog called *National Defense*: "Agencies like the Defense and State Departments are struggling to keep up with threats such as the proliferation of armed robots and alarmingly sophisticated cyber spying technologies," it read.[5] "We don't know how vulnerable we are in some cases. It is constantly moving," DoD undersecretary for acquisition Frank Kendall told the publication. Kendall's position infuriates Leitner, who believes that many 1990s-era bureaucrats who contributed to the situation are now sitting around scratching their heads, wondering how we got here. (Apparently Kendall has real reason to be concerned. Since IBM's computer-server business was sold to the Chinese-owned Lonovo Group

Ltd. in 2014,[6] the Pentagon has been buying computers and servers made in China by a Chinese-owned company.) What could possibly go wrong?

When I turned in my first draft of this book, my excellent editor at Bloomsbury Publishing, Anton Mueller, fell into a gaping hole that I'd left at the end. The book, he said, had so much going on, there needed to be a clear wrap-up of how our country arrived at the precarious situation we're in. I had plenty of Jim going on about this throughout the book. But it didn't feel right having him do the summation. I didn't want to do it myself either. Thinking about making it *my* commentary felt too overwhelming. I'm braced for some significant backlash, but making myself the expert on the decline of Western civilization was too much. I needed backup.

Then, in early 2016, the clouds parted and a rainbow appeared in the form of a retired general named John Adams. John (yes, he lets me call him John, which still feels weird) walked into the story at the last minute and gave this book an unshakable credibility. And to think I almost missed him. I found out that General Adams was on board because Jim copied me on an e-mail to J. J. Brown: "When we bring the general to go see House Armed Services . . ." *Excuse me, WHAT? A general? JIIIIIIIMMM!* Lo and behold, that week in Washington I laid eyes on the most beautiful sight: a tall, eloquent retired army general with a thirty-five-year military career and a library of research on our country's reckless negligence of our national security. I'm not sure who was happier: Jim Kennedy to have General Adams by his side, or me to have him validating everything this book is about. "I think a lot of times your duty as a military officer is to advise on what's good for the country, what's good for national security," he says. "We should train for the worst and do our very best to prevent the worst from happening." In Adams's prevention category: *Do not let China control our weapons systems.*

An important idea for sure, since in 2016 Washington had to respond to the fact that the Chinese were not only building their own islands in the strategic South China Sea but landing military aircraft there. Admiral Harry Harris, a top naval commander in the Pacific, was understandably up in arms and pushing for a tough U.S. response (perhaps he is not yet aware that all of his systems rely on Chinese parts). The *Navy Times* reported (in an article that's since been removed), that the White House responded by telling military leaders: *On the topic of China, zip it.*[7]

From my man-on-the-street perspective, it was stunning to hear General Adams's opinion of American foreign policy since the early 1990s. Adams is a harsh critic of America's political leaders, who, he says, through lies and deceit, dragged our country into an unimaginable quagmire in the Middle East. "It's pretty easy to find American young people who will go into combat," he says. "That's not the challenge. The challenge is trying to prevent politicians from wasting their lives." Adams thinks that Americans are finally waking up to being duped for the last thirty years, and he sees a revolution brewing. "Those who benefited from the successful betrayal of the people are terrified at losing control. I think they realize that they're actually going to be held accountable for it."

As he waits for any word out of Washington in the winter of 2017, John Kutch is guardedly optimistic. At the same time, he's consumed by fear. "We're more or less at the mercy of a handful of people who may or may not decide to solve this issue in the matter of few pen strokes," he says. John knows that the dream of the co-op coming to fruition could quickly be followed by a dismal nightmare—something he and Jim rarely discuss: the idea that once the project is greenlit, Jim Kennedy and John Kutsch will not be invited to the game.

"I'm still unbelievably fearful of the idea that the little kingfisher bird who dives into the water and works real hard and pulls out a fish and along comes the big eagle and rips the fish out of his face. You can imagine after ten years of work, just to hand it to somebody else who will take it and bastardize it like Molycorp and suck it dry and not solve anything. Then that's that. I'd be horrified by that. But I could only imagine what Kennedy would think. He would probably go mental." Still, John is hopeful enough that one day Western countries will have access to thorium that he's organized the eighth annual Thorium Energy Alliance meeting for August 2017 in Jim's hometown of St. Louis.

"Jim is in a funny position," says the understating J. J. Brown, "because he funds this on his own, along with John, and Jim's wife isn't a big fan of the resources and time and stress. Jim keeps saying, 'I'm done, I'm done, I'm done . . .' but every time there's new hope, he can't let it go." J. J. understands. "The thing is, he's right. This is a really serious issue, and his is the right answer. Nobody has any other answers. He knows that, and he can't ignore that."

On a beautiful June afternoon in 2016, the very last day before the Senate's defense act deadline, Jim and J. J. were still lobbying Senate staffers in Washington, D.C. I asked J. J. why he believes that Jim keeps coming back. Why is he still here when the odds are so low? How can he go on when he's been beaten down so many times? On a shady park bench behind the Hart Senate Office Building, the devout Mormon became philosophical.

J. J.: Maybe I'm getting a little bit biblical. I think that this is what Jim's *supposed* to be doing. I don't think he's self-selected. He's the guy to do this, and I really believe that. If you look at prophets in the Old Testament, first of all, they are never liked by the people they prophesy to.

JIM: That's *check one.*

J. J.: Isaiah gave incredibly detailed projections about what was going to happen to Israel, including the name of the king who was going to take over Israel. But in Israel, if you were a prophet, you would testify, and they would write down your prophesies. Twenty years later, they decided if you were a true prophet based on whether it came true. Isaiah, who ended up being the most accurate prophet, was sawed in half with a rusty saw by his own people.

JIM: I can see that coming. [*laughing*] It's gonna be Nina.

J. J.: Being a prophet doesn't mean that people listen. Sometimes it means they just need to get the message. Then it's on them. Jim said that he's here because he feels it's his job to get the message to the people who can do something.

JIM: But in your example, Israel is destroyed, right?

J. J.: Yes. Prophets aren't always successful in getting people to act, but they're still doing their jobs. They don't have a fun job, and they're rarely recognized as a prophet at the time. It's kind of silly to talk about this biblically, but to be honest, that's what we're doing:

prophesying. We're telling people what the truth is and what's going to happen. They don't even disagree.

JIM: But it's not theologically based. It's technology based. And if we can't get back in the game, I prophesize—this is a radical prediction based on *science*—that we will fall out of all these industries—that we will fall out of relevance as an economic power.

As dismal as things were, J. J. was still hopeful. "I really think long-term this is going happen," he says. "I really don't think the Pentagon is going to be able to handle it any other way."

In case you missed it, there is a love story in all of this that is pretty amazing. There is nothing that Jim does that he doesn't cross-check with Nina first. She's really put herself out on a limb supporting his endeavors for a very long time. There have been rough times, for sure. But when you see them together, you know right away that they are a team. They laugh and joke and tease each other. He exasperates her. She grounds him. "We've been together for twenty-five years," Nina says. "I still get nervous and wake up and think, 'Oh my god, what are we doing?' But I always come back to the fact that I trust Jim. I know he needs to be on track sometimes, and I need to help with the more practical details like packing his dress slacks for high-level meetings. But ultimately, I do know that he's right." For years, family and friends have told Nina to make Jim stop, make him quit this crazy fight, make him settle down. "I tell them, 'Then what's Jim going to do? He's just going to find another big project that he'll delve into, and we'll be in the same spot pursuing some other passion.'"

Jim is thankful Nina understands him. "I think that's what everyone loses in life—they actually don't realize that life is a journey. Taking risks and submersing yourself into something completely is the point of it. Most people say, I just want to rinse myself of it all, isolate myself and sit on a beach. Then you're into your third drink with an umbrella and you're losing your mind. You have to actually engage in the journey and enjoy it. Instead, it should be 'What can I do with this beach?'

"I'm pitching a tent," he told me. "I will be here to the finish line."

Acknowledgments

No work of nonfiction is created independently, and it is only because of generous help from an enormous cast of characters that I put the pieces of this story together. Agent-for-life Peter McGuigan—thank you for encouraging me to dust off an old idea at just the right time and shepherding it with the same finesse and care you do with all my books. It was with an incredible team of talent at Bloomsbury Publishing that this project came to life: To my excellent editor Anton Mueller—my sincerest gratitude for having the confidence in me to flesh out this complex topic and crafting my work into something so much richer. A million thanks to publishing director George Gibson, editorial director Nancy Miller, Patti Ratchford and the art department, production editor Jenna Dutton, editor Rachel Mannheimer, the sales team, and Anton's assistant, Grace McNamee. Super props to Miranda Ottewell's copyediting skill, and endless appreciation for the superior legal advice of Alan J. Kaufman.

Barbara Bruce—my very first editor from the time I wrote my first words—you are the best and most supportive mother a kid could ever ask for. Thank you for countless hours poring over drafts, tweaking grammar, and catching typos. Kay Scarlett, my college roommate and a geophysics genius, took on part-time editing for me, unexpectedly revealing some serious science editing chops. Kay held me accountable on the science side and didn't let me gloss over explanations I occasionally tried to pass off because they were "too hard." Sara Natividad, I am so lucky to have had your assistance in the research and writing of the historical chapter on Pea Ridge. Your lovely words remain in the final manuscript. Emilie Lorditch, thanks to you for all your quick responses to my random queries on this or that physicist or invention. Karin Hayes, you're still right here with me and a constant source of excellent feedback on all topics relevant to life as a freelancing vagabond. My brother, Tim, an avid reader of every non-fiction genre and student of life, thank you so much for your comments and feedback. Sheri Fink, you were the very first champion of this story, and I've never forgotten your encouragement. Holly Hodder, there will never be a book I write that I don't feel like you're the reason I get to do this amazing work. To brilliant photographer friend and fellow Annapolis mommy Heather Crowder, thank you for the wonderful author photo that reflects the intensity I feel about the grave topic of this book.

My crew at home: my super awesome daughter, Evelyn Dahl, spent her eleventh year delighted by bedtime stories of Jim Kennedy's shenanigans as a boy growing up in Missouri. Evelyn told me what was boring and what should stay in. Dawn Lamonica is my sister who traveled on wild adventures through China, Missouri, New York, and Illinois and helped me keep track of my massive piles of papers and references. Alex Provenzano, my true love and biggest supporter, thank you for always having your arms around me when I'm afraid and my confidence is tanking. Joe Bruce—Dad, you are my hero. Thank you for being by my side and giving me the courage to find the answers—no matter how hard the journey. Tereaz Scholze, you are my teacher and mentor. Because of you, everything that I create, I put into the perspective of a vast universe. That calms me. Ludia Sarmast, my sister-wife, your care was essential so many days. I love you all.

I have undying gratitude to Howard Schubiner, MD, whose book *Unlearn Your Pain* relieved me from chronic migraines that plagued my life and got progressively worse as the intensity of my research on this book escalated. Today, I am migraine-free. Thank you, Howard.

A shout-out must go to others who eased my pain without even knowing. Amy Poehler, your autobiography came at just the right time. Being a mom is hard. Writing a book is hard. Thanks for validating that for me and making me laugh out loud. Laura Hillenbrand, thank you for *Seabiscuit*—the Audible version got me through a rough patch. Evelyn and I listened to your warm and artistic story night after night. When I realized there was a big Wall Street component to this book, I froze in my tracks. Then I found Michael Lewis. My library is now full of dog-eared Michael

Lewis titles. His easy style gave me confidence that I could figure out how to make it work. When I felt dismal about the state of affairs in America, there was Tina Fey and *30 Rock* reruns to put me to bed at night.

If you've already read this book, you know well how Jim Kennedy opened his life and gave me full access to the complicated inner workings of his brain and the deep caverns of his soul. He is a dream come true for any author. He's exciting. He's super smart. He's funny. He tells a great story—always using the present tense to keep you on the edge of your seat. He made my work so much easier at times—other times, so much more difficult. There is no stone that Jim doesn't overturn, and he forced me to look under each one. Nina Abboud was truly amazing and helpful throughout. There is no one who doesn't love the Kennedy-Abboud collaboration. It is Yin and Yang at its finest. Thanks also to Nina's parents, Nabil and RoseMarie, who are lovely and were open and candid about having Jim Kennedy as a son-in-law. Thanks especially to Dravin and Shale, who generously shared their Halloween candy so I had something to eat after a long day with Jim (food is rarely on Jim's to-do list).

I must call out a few brilliant souls who helped make this story a solid piece of political and scientific discourse. When I pitched the idea for this book to Pulitzer Prize–winning author David Cay Johnston (who wrote about rare earths in his bestseller *Free Lunch*), his encouragement propelled my initial digging-in. Brad Van Gosen of the U.S. Geological Survey was my go-to man for all things geoscience. When almost every other geologist I came across had ties to the rare earth industry, Brad's solid perspective was invaluable. When I found Magnequench inventor John Croat in the online white pages, it was like finding the Hope Diamond of late-twentieth-century science. I'm so honored to be able to tell his story, and I cannot thank him enough for his thoughtful suggestions, edits, and help unraveling the science of rare earth magnetism. Peter Leitner was inside the sinking ship of Pentagon procurement in the 1990s, screaming at the top of his lungs when no one else was paying attention. Leitner and others who risk their careers to do the right thing have my utmost respect.

At times over the course of writing this book, I worried about future attacks on Jim's credentials by industry and Pentagon insiders, even though all of his predictions came true. When Jim casually mentioned "the

General" in a 2016 e-mail about a lobbying trip to Washington, my heart swelled. With General John Adams echoing Jim, John, and J. J. Brown in their call for Washington and the Pentagon to get their heads out of the sand, "the General" became my book's knight in shining armor. The lives of other fascinating characters fill the pages of this book: Dick Engel, Cheryl Seeger, J. J. Brown, Stan Short, and John Kutsch. A million thanks.

An enormous debt of gratitude to the following who gave up their time to help me work it all out:

Ben Auld, Mining Plus
Jack Blum, attorney
Fred Brewer, Delco Remy, retired
Chen Kun, Chinese Academy of Science
Andy Childers, Doe Run Company
Roderick Eggert, Colorado School of Mines
Richard Engel, Oak Ridge National Laboratory, retired
Charles Forsberg, Massachusetts Institute of Technology
Tom Gallagher, Pea Ridge
Uri Gat, Oak Ridge National Laboratory, retired
Jess Gehin, Oak Ridge National Laboratory
Sherrell Green, Oak Ridge National Laboratory
Karl Gschneidner, Ames Laboratory
James Hansen, NASA, retired
Mark Haselhorst, Jim's friend
David Holcomb, Oak Ridge National Laboratory
John E. Kelly, Department of Energy
Elizabeth Kennedy, Jim's mother
Mary Kennedy, Jim's sister
Vince Lackowski, Thorium Energy Alliance
Teri Luna, United Steel Workers Representative, formerly Magnequench
Franklin McCallie, Kirkwood H.S., retired
Anne E. McCafferty, U.S. Geological Survey
Bruce Moyer, Oak Ridge National Laboratory

Cliff Newman, U.S. Special Forces, retired
Andrew Nyce, former president and owner of Gorham Advanced Materials Institute
Robert Orr, attorney
Cecil Parks, Oak Ridge National Lab
Ed Richardson, Thomas & Skinner, Inc.
William Ridley, U.S. Geological Survey
Masato Sagawa, formerly Sumitomo Corporation
Tom Schilly, mayor, Crystal City, Missouri
Cheryl Seeger, Missouri Geological Survey
Dan Shawe, U.S. Geological Survey, retired
David Summers, Curators' Professor Emeritus, Missouri University of Science of Technology
Tetsuichi Takagi, Geological Survey of Japan
Rich Todaro, Jim's friend and Kennedy Capital, retired
Larry Tucker, Pea Ridge mining engineer
Ron Walli, Oak Ridge National Laboratory
Andy Worrall, Oak Ridge National Laboratory
John Wright, former president, Big River Minerals
Volker Zepf, Universität Augsburg
Matthew D. Zolnowski, Jeff Green & Company

Notes

1: Oak Ridge, Tennessee

1 Xu, "China Announces Thorium Molten Salt Reactor."
2 Johnson, *City behind a Fence*, 169.
3 Weinberg, *First Nuclear Era*.

2: Inconvenient Ideas

1 Hewlett, *Atoms for Peace and War*, 1.
2 Eisenhower, *White House Years*.
3 Hewlett, *Atoms for Peace and War*.
4 Dwight D. Eisenhower, "Atoms for Peace."
5 Weinberg, *First Nuclear Era*, 45–46.
6 Regan, "Road Trip."
7 Center for Oak Ridge Oral History, "Engel, Richard."
8 Weinberg, *First Nuclear Era*, 95–108.
9 Salts are compounds formed by one metal and one nonmetal element in their ionic states. Common table salt is composed of a positively charged sodium ion and a negatively charged chlorine atom that form an "ionic bond." Some salts melt at relatively low temperatures to become clear liquid with the consistency of water.
10 Center for Oak Ridge Oral History, "Haubenreich, Paul."
11 Haubenreich and Engel, "Experience with the Molten-Salt Reactor Experiment."
12 Weinberg, *First Nuclear Era*.
13 Weinberg, "Global Effects of Man's Production of Energy."

4: The Biggest Puzzle

1 Pigs are formed by pouring molten metal into a common reservoir (the mother pig); the metal then runs into smaller oval compartments (the suckling pigs) to solidify.
2 Houck, *History of Missouri*, 368.
3 Burford, "Underground Treasures," 5.
4 St. Joe Minerals Corporation, "History of St. Joe," 1–34.

5 St. Joe Minerals Corporation, "Mangled Iron-Er," 1959.
6 St. Joe Minerals Corporation, "Mangled Iron-Er," December 14, 1959.
7 St. Joe Minerals Corporation, "Mangled Iron-Er," August 28, 1961.
8 Woodside, "Old Miners Remember Pea Ridge."
9 Klott, "Seagram vs. St. Joe."
10 Beirne, "Big River Acquires Pea Ridge."

6: A Better Mousetrap

1 Gschneidner, *Rare Earths*, 41.
2 Clark, "High-Field Magnetization."
3 Groves, "Not Like the Old Days."
4 Japan Patent Office, "Tokushichi Mishima MK Magnetic Steel."
5 Senno, Tawara, and Hirota, "Samarium Cobalt Alloys."
6 U.S. Patent and Trademark Office, "America Invents Act."
7 History.com, "Charles Kettering Receives Patent."
8 Boyd, *Charles F. Kettering*, 82.
9 Delco Remy Group, "Better Mousetrap."
10 Ibid.
11 Mooney, *Republican War on Science*, 51–54.

9: Indigenous Innovation

1 Feigenbaum, *China's Techno-Warriors*, 229.
2 Ibid, 90–94.
3 Ibid, 141–143.
4 Kuhn, *Man Who Changed China*, 91.
5 Rene, *China's Sent-Down Generation*, xi.
6 Encyclopedia of World Biography, "Deng Xiaoping."
7 Chinese Academy of Sciences, "Jiang Mianheng."
8 Singer, "Jiang's Son Helps Bridge 2 Nations."
9 Ibid.
10 Ibid.
11 Beijing, "Report—November 1996."
12 Liu et al., "Chinese Views of Future Warfare."
13 Ibid.
14 Li, *China's Leaders*, 196.
15 Beijing, "Report—November 1996."
16 Cox, H.R. Report No. 105–851.
17 Milholin, "Trading with the Enemy."

11: The Acquisition

1 "GM to Sell Magnequench International."

2 Tkacik, *Magnequench*.
3 "GM to Sell Magnequench International."
4 "When Deng Xiaoping's Southern Tour Said."
5 Erler, "Kane Magnetics."
6 Faison, "Analysis."
7 Kifner, "Jiang Visit."
8 Purdum, "Jiang Does Business."
9 Cox, H.R. Report No. 105–851.
10 Risen and Gerth, "U.S. Is Said to Have Known of China Spy Link."
11 Marquis, "Some Lawmakers Urging."
12 C-SPAN, "2003 State of the Union."
13 Bayh and Visclosky, "REO-Magnequench2003."
14 Evans, "Ask the White House."
15 Leitner, *Decontrolling Strategic Technology*.

12: The Cowboy

1 Tritto, "$1 Billion Plan."

15: The Vision

1 AmericanTowns.com, "City Of Crystal City."
2 City of Crystal City, "History."
3 Aeppel, "Jobs Unwanted."
4 Garrison, "Sunshine Law Case Dismissed."
5 Kingsley, "Smelter Proposal."

16: War

1 Kennedy, Bio-breeder system.
2 McBride and Taylor, "Update 1."
3 Riseborough, "Glencore, Wings Study."
4 *Alberici Constructors, Inc.*, "Notice of Lis Pendens."
5 Ibid.
6 Thornsen, "Crystal City Smelter Plans."
7 Brown, "Alberici Partners with Canadian Firm."
8 "MFC Industrial Acquires Distressed Iron Ore Mine."
9 Scott, "Kennedy Cut Out."
10 Bartle, "MFC Industrial, LTD., Alberici Active."
11 Bartle, "Pea Ridge Mine Project."
12 Ibid.
13 "Iron Ore Monthly Price."

17: A Meeting of Minds

1 Hof, "Lessons from Sematech."
2 U.S. Nuclear Regulatory Commission, "Part 75—Safeguards on Nuclear Material."
3 Van Gosen et al., *Deposit Model for Heavy-Mineral Sands.*
4 I made several attempts to find out why the threshold was set at 0.05 percent. Nuclear Regulatory Agency representatives told me that the law was written in the 1940s, and they did not have any answer as to why that value was chosen.

18: Rough Waters

1 Fauna, "Chinese Fishing Boat Captain Arrested."
2 Manyin, "Senkaku Islands Dispute."
3 "Truth of the Senkaku Incident 4."
4 "Chinese Fishing Boat."
5 Fauna, "Chinese Fishing Boat Captain Arrested."
6 Bloomberg, "China Denies Japan Rare-Earth Ban."
7 Bradsher, "China Blocks Vital Exports to Japan."
8 Bradsher, "China Bans Rare Earth Exports."
9 Krugman, "Rare and Foolish."
10 Johnston, "How New and Assertive?"
11 Galyen, statement, *China's Monopoly on Rare Earths.*
12 Nelson, "REO Letter to Senator Nelson." Multiple inquirers to Chai asking whether he received a response went unanswered. Sen. Nelson never took up any rare earth legislation.
13 University of Nevada–Las Vegas, "Clark County."
14 Olson et al., *Rare-Earth Mineral Deposits.*
15 Wright et al., *Mines and Mineral Deposits,* 121–123.
16 Zhi Li and Yang, "China's Rare Earth Ore Deposits."
17 Massachusetts Institute of Technology, "Critical Elements for New Technologies."
18 Zepf, *Rare Earth Elements,* 41.

19: China Inc.

1 "Guangzhou Industrial Manufacturing Zones."
2 "China's Power Families," July 10, 2012.
3 Labaton, "F.D.I.C. Sues Neil Bush."
4 Balfour and Einhorn, "A Bush in Hand."
5 Garnaut, "In Thrall."
6 Fornet, "Chinese Leader's Son."
7 U.S.-China Economic and Security Review Commission, report.
8 Browne, "Immelt on China."
9 U.S.-China Economic and Security Review Commission, report.
10 "Locke."

20: Smoke and Mirrors

1 Swallow, *Annual Reports of the Geological Survey of Missouri*, IV.
2 *Rare Earth Minerals and 21st Century Industry.*
3 Spears and Scholer, "Molycorp Chops Rare-Earth Initial Public Offering."
4 Securities and Exchange Commission, "Registration Statement."
5 Ibid.
6 Verplanck et al., *Deposit Model for Carbonatite*, 10, Table 3.
7 Emsbo et al., "Rare Earth Elements"; Long et al., *Principal Rare Earth Elements.*
8 Lynas, "J. P. Morgan Initiates Coverage of Lynas."
9 Tan, "Lynas Rare Earth Project."
10 Galyen, statement, *China's Monopoly on Rare Earths*, 31–33.
11 Kendall, Rare Earth Materials.
12 Ratnam, "Rare Earth Supplies."
13 Areddy and Hodge, "Pentagon Plays Down."
14 Grasso, "Rare Earth Elements," 1, 14.
15 Verplanck et al., *Deposit Model for Carbonatite*, 10, Table 3.
16 "CMI Factsheet."
17 Obama, "Remarks on Fair Trade."
18 Bradsher, "Merger."

21: The First to Eat a Crab

1 "China's Power Families."
2 Springut, et al., *China's_Program*, 84–85.
3 Ibid, 85–86.
4 World Nuclear Association, "Nuclear Power in China."
5 Ibid.
6 Hargraves and Moir, "Liquid Fluoride Thorium Reactors."
7 "Xu Hongjie," video.
8 "Jiang Mianheng," video.
9 U.S. Government Accountability Office, *Advanced Reactor Research*, 28.
10 Suttmeier, "Trends in US-China Science," 4.
11 Chinese Academy of Sciences and U.S. Department of Energy, "First Executive Committee Meeting."
12 U.S. Department of Energy, "FY 2015."
13 Ibid.

22: Maneuvers

1 Gardner, "Tea Party Wins Victory."
2 Freking and Loftin, "Hatch Faces Obstacle."
3 Stockman, National Rare-Earth Cooperative Act.
4 The formal procedure for introducing legislation in the House is to put a copy of the bill in a special box called the hopper on the speaker's podium.

5 U.S. Department of Energy, "FY 2015."
6 Sternberg, "Great Rare-Earth Metals Crisis."
7 Glaser, "U.S. Rare Earths Policy."
8 Gholz, *Rare Earth Elements*, 1.
9 Worstall, "Why Lynas Corp Is Struggling."

23: Zero Sum Game

1 Matich, "Molycorp Still Planning."
2 Inspector General, *Procedures to Ensure Sufficient Rare Earth Elements*, i.
3 Adams and Kurzer, "Remaking American Security," vi.
4 U.S. Government Accountability Office, *Rare Earth Materials*, 13.
5 Committee on Armed Services, "Full Committee Markup of H.R. 4909."

Epilogue

1 Lienert and Thompson, "GM Didn't Fix Deadly Ignition Switch"; Korosec, "Ten Times More Deaths."
2 Zepf, *Rare Earth Elements*.
3 Gholz, *Rare Earth Elements*, 1.
4 Miller and Zheng, "Molycorp Files for Bankruptcy Protection."
5 Erwin, "'Democratization' of Technology."
6 Tsukayama and Lamothe, "How an Email Sparked a Squabble."
7 Lamothe, "Gag Order Issued on South China Sea?"

Bibliography

Adams, John, and Paulette Kurzer. *Remaking American Security: Supply Chain Vulnerabilities & National Security Risks across the U.S. Defense Industrial Base.* Alliance for American Manufacturing, 2013. http://www.americanmanufacturing.org/research/entry/remaking-american-security.

Adams, John W., and Mortimer H. Staatz, "Rare-Earth Elements," in *United States Geological Survey Professional Paper 820*, ed. Donald A. Brobst and Walden P. Pratt (Washington, D.C.: United States Printing Office, 1973), 547–556.

Aeppel, Timothy. "Jobs Unwanted." *Wall Street Journal*, September 17, 2008. http://www.wsj.com/articles/SB122152667328940259.

Alberici Constructors, Inc., v. Upland Wings, Inc., Wings Enterprises, Inc., James Kennedy, Nina Abboud. "Notice of Lis Pendens." November 5, 2010. Print.

AmericanTowns.com. "City of Crystal City." Accessed September 22, 2015. http://www.americantowns.com/mo/crystalcity/organization/city_of_crystal_city.

Areddy, James T., and Nathan Hodge. "Pentagon Plays Down China's Rare-Earths Controls." *Wall Street Journal*, March 16, 2012. http://www.wsj.com/articles/SB10001424052702304459804577285022878180992.

Balfour, Frederik, and Bruce Einhorn. "A Bush in Hand Is Worth . . . a Lot." *BloombergView*, December 15, 2003. http://www.bloomberg.com/bw/stories/2003-12-14/a-bush-in-hand-is-worth-dot-dot-dot-a-lot.

Bartle, James B. "MFC Industrial, LTD., Alberici Active in Pea Ridge Mine Project CEO Stated, 'Project Is Full Steam Ahead.'" *Sullivan Independent*

News, April 10, 2013. http://www.mysullivannews.com/2013/04/mfc-industrial-ltd-alberici-active-in-pea-ridge-mine-project-ceo-stated-project-is-full-steam-ahead/.

———. "Pea Ridge Mine Project Still A Reality." *Sullivan Independent News*, August 12, 2014. http://www.mysullivannews.com/2014/08/pea-ridge-mine-project-still-a-reality/.

Bayh, Evan, and Peter J. Visclosky. Evan Bayh and Peter J. Visclosky to President George W. Bush, Washington, D.C., March 6, 2003.

Beirne, Mike. "Big River Acquires Pea Ridge (Big River Minerals Corp.; Pea Ridge Iron Ore Co.)." American Metal Market, July 3, 1990. http://business.highbeam.com/436402/article-1G1-9250025/big-river-acquires-pea-ridge.

Bloomberg. "China Denies Japan Rare-Earth Ban Amid Diplomatic Row." *Washington Post*, September 23, 2010. http://www.washingtonpost.com/wp-dyn/content/article/2010/09/23/AR2010092300277.html.

Boyd, Thomas Alvin. *Charles F. Kettering: A Biography*. Washington, D.C.: Beard Books, 2002.

Bradsher, Keith. "Amid Tension, China Blocks Vital Exports to Japan." *New York Times*, September 22, 2010. http://www.nytimes.com/2010/09/23/business/global/23rare.html.

———. "China Bans Rare Earth Exports to Japan amid Tension." CNBC online. Accessed October 25, 2015. http://www.cnbc.com/id/39318826.

———. "Merger Combines a Rare Earth Mining Firm with a Processor." *New York Times*, March 8, 2012. http://www.nytimes.com/2012/03/09/business/global/merger-combines-a-rare-earth-mining-firm-with-a-processor.html.

Brown, Lisa. "Overland-Based Alberici Partners with Canadian Firm to Buy Mine." Stltoday.com. Accessed May 16, 2016. http://www.stltoday.com/business/local/overland-based-alberici-partners-with-canadian-firm-to-buy-mine/article_5ca0970a-36f2-11e1-8bdf-001a4bcf6878.html.

Brown, Lisa, and Leah Thorsen. "Wings, Swiss Giant Glencore Strike Deal to Develop Sullivan Mine." *St. Louis Post-Dispatch*, October 20, 2010. http://www.stltoday.com/business/article_97e73974-285d-5f3e-ad51-d6609333352b.html.

Browne, Andrew. "Immelt on China: They Won't Let Us Win." *WSJ China Realtime*, July 2, 2010. http://blogs.wsj.com/chinarealtime/2010/07/02/immelt-on-china-they-wont-let-us-win/.

Burford, Jo. "Underground Treasures." In *Official Manual State of Missouri, 1977–1978*, 1–33. Jefferson City: Von Hoffmann Press, n.d.

C-SPAN. "2003 State of the Union." Accessed October 25, 2015. http://www.c-span.org/video/?174799-2/2003-state-union.

Center for Oak Ridge Oral History. "Engel, Richard." Oral history interview by Keith McDaniel, December 8, 2011. http://cdm16107.contentdm.oclc.org/cdm/ref/collection/p15388coll1/id/190.

———. "Haubenreich, Paul." Oral history interview by Stephen H. Stow, March 13, 2003. http://cdm16107.contentdm.oclc.org/cdm/ref/collection/p15388coll1/id/259.

Chai, Bruce. Bruce Chai to Senator Bill Nelson, Washington, D.C., March 4, 2011.

"China's Power Families." *Financial Times*, July 10, 2012. http://www.ft.com/intl/cms/s/2/6b983f7a-ca9e-11e1-8872-00144feabdc0.html#axzz3pXVj5wLy.

Chinese Academy of Sciences and U.S. Department of Energy. *Cooperation In Nuclear Energy Sciences And Technologies: First Executive Committee Meeting*. Summary Meeting Report. Shanghai, China. October 22, 2012. Print.

Chinese Academy of Sciences. "Jiang Mianheng." Profile page, 2009. http://sourcedb.cas.cn/sourcedb_sim_cas/en/expert/200907/t20090701_1884392.html.

"Chinese Fishing Boat Rams the Japan Coast Guard's Ship." YouTube video, posted by "boco526," November 4, 2010. https://www.youtube.com/watch?v=lvo31K_lV4I.

City of Crystal City, Missouri. "History." Accessed September 22, 2015. http://crystalcitymo.org/history/index.html.

Clark, Arthur E. "High-Field Magnetization and Coercivity of Amorphous Rare-Earth-Fe2 Alloys." *Applied Physics Letters* 23, no. 11 (1973): 642–44.

Critical Minerals Institute. "CMI Factsheet." Accessed October 25, 2015. https://cmi.ameslab.gov/materials/factsheet.

Committee on Armed Services. Full Committee Markup of H.R. 4909—National Defense Authorization Act for Fiscal Year 2017. April 27, 2016.https://armedservices.house.gov/legislation/markups/FChr-4909-national-defense-authorization-act-fiscal-year-2017-5.

Cox, Christopher. H.R. Report No. 105-851 (1999) for the Select Committee on US National Security and Military/Commercial Concerns with the People's Republic of China. https://www.gpo.gov/fdsys/pkg/GPO-CRPT-105hrpt851/pdf/GPO-CRPT-105hrpt851.pdf.

Danfoss. "Danfoss North America President Testifies on Rare Earths before Congressional Committee." Press release, September 26, 2011.

Delco Remy Group, General Motors. "We Built a Better Mousetrap." *Clan*, April 1985.

"Deng Xiaoping Biography." *Encyclopedia of World Biography*. Accessed June 1, 2015. http://www.notablebiographies.com/De-Du/Deng-Xiaoping.html.

Eisenhower, Dwight D. "Atoms for Peace." Text of speech presented at the 470th Plenary Meeting of the United Nations General Assembly, India, December 8, 1953. https://www.iaea.org/about/history/atoms-for-peace-speech.

———. *The White House Years*. Garden City, NY: Doubleday, 1963.

"Election Financiers." *Finland Today*, May 21, 2008. https://finlandtoday.wordpress.com/2008/05/21/election-financiers-2/.

Embassy of the United States. "Report—November 1996." November 27, 2012. http://beijing.usembassy-china.org.cn/report1196education.html.

Emsbo, Poul, Patrick I. McLaughlin, George N. Breit, Edward A. du Bray, and Alan E. Koenig. "Rare Earth Elements in Sedimentary Phosphate Deposits: Solution to the Global REE Crisis?" *Gondwana Research* 27, no. 2 (February 2015): 776–85. doi:10.1016/j.gr.2014.10.008.

Erler, Susan. "Kane Magnetics Buys Former Magnequench Plant." Nwi.com, May 26, 2004. http://www.nwitimes.com/business/local/kane-magnetics-buys-former-magnequench-plant/article_8ce665ff-3f1a-575f-adc4-8992fee84c7f.html.

Erwin, Sandra M. "'Democratization' of Technology Rattles U.S. National Security Agencies." *National Defense*, March 2, 2016. http://www.nationaldefensemagazine.org/blog/lists/posts/post.aspx?ID=2108.

Evans, Don. "Ask the White House." Online forum, June 26, 2003. http://georgewbush-whitehouse.archives.gov/ask/text/20030626.html.

Faison, Seth. "Analysis: U.S. Trip Is Everything Jiang Expected." *New York Times*, November 3, 1997. https://partners.nytimes.com/library/world/110397us-china-assess.html.

Fauna. "Chinese Fishing Boat Captain Arrested by Japanese, Reactions." *chinaSMACK*, September 10, 2010. http://www.chinasmack.com/2010/stories/chinese-fishing-boat-captain-arrested-by-japanese-reactions.html.

Feigenbaum, Evan A. *China's Techno-Warriors: National Security and Strategic Competition from the Nuclear to the Information Age*. Stanford, Calif.: Stanford University Press, 2003.

Forney, Matt. "Chinese Leader's Son Builds Telecommunications Empire." *Wall Street Journal*, November 1, 1999. http://www.wsj.com/articles/SB941414623101926633.

Freking, Kevin, and Josh Loftin. "Hatch Faces Obstacle Despite Reaching out to Base." *Huffington Post*, July 14, 2011. http://www.huffingtonpost.com/huff-wires/20110614/us-hatch-tea-party-outreach/.

Gardner, Amy. "Tea Party Wins Victory in Utah as Incumbent GOP Senator Loses Bid for Nomination." *Washington Post*, May 9, 2010. http://www.washingtonpost.com/wp-dyn/content/article/2010/05/08/AR2010050803430.html.

Garnaut, John. "In Thrall of the Empire of the Sons." *Sydney Morning Herald*, May 6, 2012. http://www.smh.com.au/world/in-thrall-of-the-empire-of-the-sons-20120525-1za2d.html.

Garrison, Chad. "Sunshine Law Case Dismissed; Judge Cites Anonymity of Online Commenters." *Riverfront Times*, December 2, 2009. http://www.riverfronttimes.com/newsblog/2009/12/02/sunshine-law-case-dismissed-judge-cites-anonymity-of-online-commenters.

Geological Survey Professional Paper. U.S. Government Printing Office, 1973.

Gholz, Eugene. *Rare Earth Elements and National Security: A CFR Energy Report*. New York: Council on Foreign Relations, 2014. http://www.cfr.org/energy-and-environment/rare-earth-elements-national-security/p33632.

Glaser, William R. "U.S. Rare Earths Policy: Digging Out of the Rare Earths Quandary." *Luce.nt: A Journal of National Security Studies*, Fall 2014. https://www.usnwc.edu/Publications/-Luce-nt-/Archives/2014/Special-Edition/Glaser_RareearthspolicyFormatted-and-Watermarked.aspx.

"GM to Sell Magnequench International." PR Newswire, June 28, 1995. http://www.thefreelibrary.com/GM+TO+SELL+MAGNEQUENCH+INTERNATIONAL-a017155332.

Grasso, Valerie B. *Rare Earth Elements in National Defense: Background, Oversight Issues, and Options for Congress*. CRS Report No. R41744. Washington, DC: Congressional Research Service, 2013. http://oai.dtic.mil/oai/oai?verb=getRecord&metadataPrefix=html&identifier=ADA590410.

Groves, Martha. "Not Like the Old Days: GM, Union No Longer Wield the Same Clout." *Los Angeles Times*, March 22, 1996. http://articles.latimes.com/1996-03-22/business/fi-49997_1_union-members.

Gschneidner, Karl A. Jr. *Rare Earths: The Fraternal Fifteen*. Oak Ridge, TN: U.S. Atomic Energy Commission, 1966. http://www.osti.gov/includes/opennet/includes/Understanding%20the%20Atom/Rare%20Earth%20The%20Fraternal%20Fifteen.pdf.

"Guangzhou Industrial Manufacturing Zones—an Overview." Accessed October 25, 2015. http://www.excelguangzhou.com/manufacturingzones.html.

Hargraves, Robert, and Ralph Moir. "Liquid Fluoride Thorium Reactors." *American Scientist* 98, no. 4 (2010): 304–13.

Haubenreich, Paul N., and J. R. Engel. "Experience with the Molten-Salt Reactor Experiment." *Nuclear Technology* 8, no. 2 (1970): 118–36.

Hewlett, Richard G. *Atoms for Peace and War, 1953–1961: Eisenhower and the Atomic Energy Commission*. Berkeley: University of California Press, 1989.

History.com. "Charles Kettering Receives Patent for Electric Self-Starter." Accessed October 23, 2015. http://www.history.com/this-day-in-history/charles-kettering-receives-patent-for-electric-self-starter.

Hof, Robert D. "Lessons from Sematech." *MIT Technology Review*, July 25, 2011. http://www.technologyreview.com/news/424786/lessons-from-sematech/.

Houck, Louis. *A History of Missouri from the Earliest Explorations and Settlements until the Admission of the State into the Union*. Chicago: R. R. Donnelley & Sons, 1908.

Inspector General, US Department of Defense. *Procedures to Ensure Sufficient Rare Earth Elements for the Defense Industrial Base Need Improvement*. Report No. DODIG-2014-091, July 3, 2014. http://www.dodig.mil/pubs/documents/DODIG-2014-091.pdf.

"Iron Ore Monthly Price." IndexMundi. Accessed October 26, 2015. http://www.indexmundi.com/commodities/?commodity=iron-ore&months=60.

Japan Patent Office. "Tokushichi Mishima MK Magnetic Steel." October 7, 2002. http://www.jpo.go.jp/seido_e/rekishi_e/tokushi_mishima.htm.

"Jiang Mianheng—Why Nuclear Power in China? Thorium & the Energy Outlook of China @ ThEC12." YouTube video, posted by "gordonmcdowell," November 24, 2013. https://www.youtube.com/watch?v=iLX8jCKL9I4.

Johnson, Charles W. *City behind a Fence: Oak Ridge, Tennessee, 1942–1946*. Knoxville: University of Tennessee Press, 1981.

Johnston, Alastair Iain. "How New and Assertive Is China's New Assertiveness?" *International Security* 37, no. 4 (2013): 7–48.

Kendall, Frank. Report to Congress: Rare Earth Materials in Defense Applications. 112th Congress (2012).

Kennedy, James C. Bio-breeder system for biomass production. U.S. Patent 20090221057 A1, filed February 28, 2008, and issued September 3, 2009. http://www.google.com/patents/US20090221057.

Kifner, John. "The Jiang Visit: The Overview." *New York Times*, November 1, 1997. http://www.nytimes.com/1997/11/01/world/the-jiang-visit-the-overview-mao-s-heir-finds-path-wall-street.html.

Kingsley, Cathy. "Alberici Constructors Case Moves Forward." *Missouri Lawyers Media*, February 17, 2011. http://molawyersmedia.com/2011/02/17/alberici-constructors-case-moves-forward/.

———. "Smelter Proposal Gets Last Public Hearing before Vote." *Missouri Lawyers Weekly*, March 9, 2009. http://molawyersmedia.com/2009/03/09/smelter-proposal-gets-last-public-hearing-before-vote/.

Klott, Gary. "Seagram vs. St. Joe." UPI, March 21, 1981. http://www.upi.com/Archives/1981/03/21/Seagram-vs-St-Joe/2686353998800/.

Korosec, Kirsten. "Ten Times More Deaths Linked to Faulty Switch than GM First Reported." *Fortune*, August 24, 2014. http://fortune.com/2015/08/24/feinberg-gm-faulty-ignition-switch/.

Krugman, Paul. "Rare and Foolish." *New York Times*, October 17, 2010. http://www.nytimes.com/2010/10/18/opinion/18krugman.html.

Kuhn, Robert Lawrence. *The Man Who Changed China: The Life and Legacy of Jiang Zemin*. New York: Crown, 2004.

Labaton, Stephen. "F.D.I.C. Sues Neil Bush and Others at Silverado." *New York Times*, September 22, 1990. http://www.nytimes.com/1990/09/22/business/fdic-sues-neil-bush-and-others-at-silverado.html.

Lamothe, Dan. "Gag Order Issued on South China Sea? Pentagon and Top Admiral Say No Way." *Washington Post*, April 7, 2016. https://www.washingtonpost.com/news/checkpoint/wp/2016/04/07/gag-order-issued-on-south-china-sea-pentagon-and-top-admiral-say-no-way/.

Leitner, Peter M. *Decontrolling Strategic Technology, 1990–1992: Creating the Military Threats of the 21st Century*. Lanham, Md.: University Press of America, 1995.

Li, Cheng. *China's Leaders: The New Generation*. Lanham, Md.: Rowman & Littlefield, 2001.

Lienert, Paul, and Marilyn Thompson. "GM Didn't Fix Deadly Ignition Switch Because It Would Have Cost $1 Per Car." *Huffington Post*, April 2, 2014. http://www.huffingtonpost.com/2014/04/02/gm-ignition-switch-dollar-per-car_n_5075680.html.

Liu, Huaqing, Huan Yang, Jianguo Wu, Benchan Chen, Henggao Ding, Wuzheng Zi, Chengyu Yang, Zhenhuan Shun, et al. "Chinese Views of Future Warfare, Part Three: Modernizing for Local Defense Modernization in Historical Perspective." Institute for National Strategic Studies. (First published in *China Military Science*, Winter 1994.) Accessed November 3, 2015. http://www.au.af.mil/au/awc/awcgate/ndu/chinview/chinapt3.html.

"Locke: China's Policies May Scare Off U.S. Firms." Newsmax, January 29, 2010. http://www.newsmax.com/Finance/InvestingAnalysis/gary-locke-china-policies/2010/01/29/id/348429/.

"London-Based Oligarchs Face Cash Probe." *Independent*, August 7, 2004. http://www.independent.co.uk/news/business/news/london-based-oligarchs-face-cash-probe-5356140.html.

Long, Keith R., Bradley S. Van Gosen, Nora K. Foley, and Daniel Cordier. *The Principal Rare Earth Elements Deposits of the United States: A Summary of Domestic Deposits and a Global Perspective.* US Geological Survey Scientific Investigations Report 2010–5220. Reston, VA: US Geological Survey, 2010. http://pubs.usgs.gov/sir/2010/5220/.

Lynas Corporation LTD. "J. P. Morgan Initiates Coverage of Lynas." Media release, June 28, 2010.

"Lynas Rare Earth Project Sparks Huge Campaign in Malaysia." Friends of the Earth Australia. Accessed October 26, 2015. http://www.foe.org.au/lynas-rare-earth-project-sparks-huge-campaign-malaysia.

Manyin, Mark E. *Senkaku (Diaoyu/Diaoyutai) Islands Dispute: U.S. Treaty Obligations.* CRS Report No. R42761. Washington, DC: Congressional Research Service, 2013.

Marquis, Christopher. "Some Lawmakers Urging U.S. to Speed Exports of Satellites." *New York Times,* July 9, 2001. http://www.nytimes.com/2001/07/09/world/some-lawmakers-urging-us-to-speed-exports-of-satellites.html.

Massachusetts Institute of Technology. *Critical Elements for New Energy Technologies.* MIT Energy Initiative Workshop Report, April 29, 2010. http://web.mit.edu/miteicomm/web/reports/critical_elements/CritElem_Report_Final.pdf.

Matich, Teresa. "Molycorp Still Planning Top Executive Bonuses." *Investing News Network,* September 29, 2015. http://investingnews.com/daily/resource-investing/critical-metals-investing/rare-earth-investing/molycorp-still-planning-top-executive-bonuses/.

Mayama, Katsuhiko. *Chinese People's Liberation Army-Reduction in Force by 500,000 and Trend of Modernization.* Bulletin. Japan National Institute for Defense Studies, 1999. http://www.nids.go.jp/english/publication/kiyo/pdf/bulletin_e1999_6.pdf.

"MFC Industrial Acquires Distressed Iron Ore Mine." PR Newswire, January 4, 2012. http://www.prnewswire.com/news-releases/mfc-industrial-acquires-distressed-iron-ore-mine-136650718.html.

McBride, Sarah, and Michael Taylor, "Update 1: Glencore Seals Rare-Earth Mine Deal with US Firm." Reuters, October 18, 2010. http://www.reuters.com/article/2010/10/18/glencore-idUSN1828412120101018.

Milholin, Gary. "Trading with the Enemy: Wisconsin Project on Nuclear Arms Control." *Commentary Magazine*, May 2002.

Miller, John W., and Anjie Zheng. "Molycorp Files for Bankruptcy Protection." *Wall Street Journal*, June 25, 2015. http://www.wsj.com/articles/SB10907564710791284872504581069270334872848.

Mooney, Chris. *The Republican War on Science*. New York: Basic Books, 2005.

Morrison, Wayne M., and Rachel Tang. *China's Rare Earth Industry and Export Regime: Economic and Trade Implications for the United States*. CRS Report No. R42510. Washington, D.C.: Congressional Research Service, 2012. https://www.fas.org/sgp/crs/row/R42510.pdf.

Obama, Barack. "Remarks by the President on Fair Trade." White House website, March 13, 2012. https://www.whitehouse.gov/photos-and-video/video/2012/03/13/president-obama-speaks-enforcing-trade-rights-china#transcript.

Olson, J. C., D. R. Shawe, L. C. Pray, and W. N. Sharp. *Rare-Earth Mineral Deposits of the Mountain Pass District San Bernardino County California.* Geological Society Professional Paper, 261. United States Government Printing Office, 1954.

Purdum, Todd S. "Jiang Does Business on Last Stop of U.S. Visit." *New York Times*, November 3, 1997. http://partners.nytimes.com/library/world/110397us-china.html.

Rare Earth Minerals and 21st Century Industry: Hearing before the Committee on Science and Technology. 111th Congress (2010). http://www.gpo.gov/fdsys/pkg/CHRG-111hhrg55844/html/CHRG-111hhrg55844.htm.

Ratnam, Gopal. "Rare Earth Supplies in U.S. to Meet Defense Needs, Pentagon Says." Bloomberg.com, April 4, 2012. http://www.bloomberg.com/news/articles/2012-04-04/rare-earth-supplies-in-u-s-to-meet-defense-needs-pentagon-says.

Regan, Raina. "Road Trip to the Secret City: Atomic History in Oak Ridge, Tennessee." National Trust for Historic Preservation, August 26, 2014. https://savingplaces.org/stories/road-trip-secret-city-atomic-history-oak-ridge-tennessee.

Rene, Helena K. *China's Sent-Down Generation: Public Administration and the Legacies of Mao's Rustication Program*. Washington, D.C.: Georgetown University Press, 2013.

U.S.-China Economic and Security Review Commission. Report to Congress, Washington, D.C.: U.S. Government Printing Office, 2010.

Riseborough, Jesse. "Glencore, Wings Study Restarting Rare Earth, Iron Mine." Bloomberg.com, October 18, 2010. http://www.bloomberg.com/news/articles/2010-10-18/glencore-international-wings-enterprises-to-restart-rare-earth-iron-mine.

Risen, James, and Jeff Gerth. "U.S. Is Said to Have Known of China Spy Link in 1995." *New York Times*, June 27, 1999. http://partners.nytimes.com/library/world/asia/062799china-nuke.html.

Scott, Peggy. "Kennedy Cut Out of Proposed Iron Smelter in Crystal City." *Leader Publications*, January 11, 2012. http://www.myleaderpaper.com/archives/kennedy-cut-out-of-proposed-iron-smelter-in-crystal-city/article_6264fc08-9cd5-5034-8cc7-c4d6077a4777.html.

Molycorp Inc., Form S-1 Filed June 21, 2010. Securities and Exchange Commission website. Accessed September 19, 2016. https://www.sec.gov/Archives/edgar/data/1489137/000095012310059414/d70469a2sv1za.htm.

Senno, Haruhumi, Yoshio Tawara, and Eiichi Hirota. "Coercive Force of New Cu-Substituted Samarium Cobalt Alloys." *Applied Physics Letters* 29 (October 1, 1976): 514. doi:10.1063/1.89149.

Serafin, Tatiana. "Emerging Market Gold." *Forbes*, March 27, 2006. http://www.forbes.com/free_forbes/2006/0327/164.html.

Singer, Rena. "Jiang's Son Helps Bridge 2 Nations: As the President Rose to Power at Home, the Son Studied At Drexel." Philly.com, October 26, 2007. http://articles.philly.com/1997-10-26/news/25540771_1_jiang-mianheng-chinese-students-drexel-university.

Spears, Lee, and Kristen Scholer. "Molycorp Chops Rare-Earth Initial Public Offering by 18% to $394 Million." Bloomberg, July 29, 2010. http://www.bloomberg.com/news/2010-07-29/molycorp-reduces-planned-rare-earth-initial-offering-18-to-394-million.html.

Springut, Micah, Stephen Schlaikjer, and David Chen, "China's_Program for Science and Technology Modernization" (CENTRA Technology Inc., January 2011).

St. Joe Minerals Corporation. "History of St. Joe." *St. Joe Headframe*, special edition, Fall 1970.

———. "The Mangled Iron-Er." *Mangled Iron-Er*, 1959.

———. "The Mangled Iron-Er." *Mangled Iron-Er*, December 14, 1959.

———. "The Mangled Iron-Er." *Mangled Iron-Er*, August 28, 1961.

Sternberg, Joseph. "How the Great Rare-Earth Metals Crisis Vanished." *Wall Street Journal*, January 8, 2014. http://www.wsj.com/articles/SB10001424052702303848104579308252845415022.

Stockman, Steve, sponsor. National Rare-Earth Cooperative Act of 2014, H.R. 4883, 113th Cong. (2013–2014). https://www.congress.gov/bill/113th-congress/house-bill/4883.

Suttmeier, Richard P. *Trends in US-China Science and Technology Cooperation: Collaborative Knowledge Production for the Twenty-First Century?* Report prepared on behalf of the U.S.-China Economic and Security Review Commission, September 11, 2014.

Swallow, George Clinton. *First and Second Annual Reports of the Geological Survey of Missouri*. Jefferson City: Missouri Geological Survey, 1855.

Tan, Lee. "The Reasons behind Motorola's Success in China: What Makes Chinese Joint Ventures a Success." *Strategic Direction* 19, no. 2 (2003): 28–30. doi:10.1108/02580540310794327.

Thornsen, Leah. "Crystal City Smelter Plans Won't Include Former Developer." STLtoday.com, February 15, 2012. http://www.stltoday.com/news/local/metro/crystal-city-smelter-plans-won-t-include-former-developer/article_cbf5f162-5c3d-53a7-8973-a09f26ccb2ff.html.

Tkacik, John J. Jr. *Magnequench: CFIUS and China's Thirst for U.S. Defense Technology*. Washington, D.C.: Heritage Foundation, 2008. http://www.heritage.org/research/reports/2008/05/magnequench-cfius-and-chinas-thirst-for-us-defense-technology.

Tritto, Christopher. "Jim Kennedy's $1 Billion Plan to Break a Chinese Monopoly." *St. Louis Business Journal*, September 26, 2010. http://www.bizjournals.com/stlouis/stories/2010/09/27/story2.html.

"Truth of the Senkaku Incident 4—Chinese Boat Ram Japan Coast Guard," YouTube video, posted by "2DfightingFTW," November 4, 2010. https://www.youtube.com/watch?v=ULHitmtvaFw.

Tsukayama, Hayley, and Dan Lamothe. "How an Email Sparked a Squabble over Chinese-Owned Lenovo's Role at Pentagon." *Washington Post*, April 22, 2016. https://www.washingtonpost.com/business/economy/how-an-email-sparked-a-squabble-over-chinese-owned-lenovos-role-at-pentagon/2016/04/22/b1cd43d8-07ca-11e6-a12f-ea5aed7958dc_story.html.

University of Nevada, Las Vegas. "Clark County." *Southern Nevada: The Boomtown Years*, 2009. University Libraries, Digital Collections. http://digital.library.unlv.edu/boomtown/counties/clark.php.

U.S. Congress. Committee on Foreign Affairs Subcommittee on Asia and the Pacific. In *China's Monopoly on Rare Earths: Implications for U.S. Foreign and Security Policy*, 2011 (112th Congress), statement of John Galyen.

U.S. Department of Energy. "FY 2015 Congressional Budget Request." March 2014.

U.S. Government Accountability Office. *Advanced Reactor Research: Report to the Chairman, Subcommittee on Energy and Water Development, Committee on Appropriations, U.S. Senate*. GAO-14-545. Washington, D.C.: US Government Accountability Office, 2014.

———. *Rare Earth Materials: Developing a Comprehensive Approach Could Help DOD Better Manage National Security Risks in the Supply Chain*. GAO-16-161. Washington, DC: U.S. Government Accountability Office, 2016. http://www.gao.gov/assets/680/675165.pdf.

U.S. Nuclear Regulatory Commission. "Part 75—Safeguards on Nuclear Material—Implementation of US-IAEA Agreement." Accessed October 26, 2015. http://www.nrc.gov/reading-rm/doc-collections/cfr/part075/full-text.html.

U.S. Patent and Trademark Office. "America Invents Act (AIA) Frequently Asked Questions." Accessed October 24, 2015. http://www.uspto.gov/patent/laws-and-regulations/america-invents-act-aia/america-invents-act-aia-frequently-asked.

Van Gosen, Bradley S., David L. Fey, Anjana K. Shah, Philip L. Verplanck, and Todd M. Hoefen. *Deposit Model for Heavy-Mineral Sands in Coastal*

Environments. U.S. Geological Survey Scientific Investigations Report 2010–5070–L, 2014. http://dx.doi.org/10.3133/sir20105070L.

Verplanck, Philip L., Bradley S. Van Gosen, Robert R. Seal, and Anne E. McCafferty. *A Deposit Model for Carbonatite and Peralkaline Intrusion-Related Rare Earth Element Deposits*. U.S. Geological Survey Scientific Investigations Report 2010-5070-J, March 3, 2014. doi: 10.3133/sir20105070J.

Weinberg, Alvin Martin. "Global Effects of Man's Production of Energy." *Science* 186, no. 4160 (1974): 205–5. doi:10.1126/science.186.4160.205.

Weinberg, Alvin Martin. *The First Nuclear Era: The Life and Times of a Technological Fixer*. New York: American Institute of Physics, 1994.

"When Deng Xiaoping's Southern Tour Said: 'The Middle East Has Oil, China Has Rare Earths.'" China Public Radio, August 16, 2007.

Woodside, Nathan. "Old Miners Remember Pea Ridge at Reunion." *Missourian*, July 2, 2010. http://www.emissourian.com/local_news/saint_clair/old-miners-remember-pea-ridge-at-reunion/article_907cd9c8-8611-11df-8599-001cc4c03286.html.

World Nuclear Association. "Nuclear Power in China." July 29, 2015. http://www.world-nuclear.org/information-library/Country-Profiles/Countries-A-F/China-Nuclear-Power.aspx.

Worstall, "Why Lynas Corp Is Struggling; The Great Rare Earth Shortage Is Truly Over," *Forbes*, August 3, 2014, http://www.forbes.com/sites/timworstall/2014/08/03/why-lynas-corp-is-struggling-the-great-rare-earth-shortage-is-truly-over/.

Wright, Lauren A., et al., *Mines and Mineral Deposits of San Bernardino County, California*, vol. 49, California Journal of Mines and Geology (San Francisco: Department of Natural Resources: Division of Mines, 1953), https://ia800302.us.archive.org/24/items/californiajourna49cali/californiajourna49cali.pdf.

"Xu Hongjie—Chinese Academy of Sciences Thorium Molten Salt Reactors Solid & Liquid Fuel @ ThEC12." YouTube video, posted by "gordonmcdowell," November 30, 2012. https://www.youtube.com/watch?v=5Chj2iJsGSU.

Xu, Qimin. "China Announces Thorium Molten Salt Reactor" (Google translation). *Wen Wei Po*, January 26, 2011.

Zepf, Volker. *Rare Earth Elements: A New Approach to the Nexus of Supply, Demand and Use: Exemplified along the Use of Neodymium in Permanent Magnets* (Springer Science & Business Media, 2013).

Zhi Li, Ling, and Xiaosheng Yang. "China's Rare Earth Ore Deposits and Beneficiation Techniques." *China Western Mining*, 2014.

Index

A note on the author

Victoria Bruce holds a master's degree in geology from the University of California, Riverside. She received the Alfred I. duPont–Columbia University Award for excellence in broadcast journalism for her first film, *The Kidnapping of Ingrid Betancourt*. Bruce also coproduced and directed *Held Hostage in Colombia*, excerpted on CBS's *60 Minutes* and broadcast on the History Channel and the Sundance Channel. Her previous books are *No Apparent Danger* and *Hostage Nation*. She lives in Anne Arundel County, Maryland.